AF598302

Seismic Reflection Interpretation

Seismic Reflection Interpretation

A. H. KLEYN

Formerly of Shell Internationale Petroleum Maatschappij BV, The Hague, The Netherlands

APPLIED SCIENCE PUBLISHERS
LONDON AND NEW YORK

APPLIED SCIENCE PUBLISHERS LTD
Ripple Road, Barking, Essex, England

Sole Distributor in the USA and Canada
ELSEVIER SCIENCE PUBLISHING CO., INC.,
52 Vanderbilt Avenue, New York, NY 10017, USA

British Library Cataloguing in Publication Data

Kleyn, A. H.
Seismic reflection interpretation.
1. Seismology 2. Signal processing
I. Title
551.2′2 QE539

ISBN 0-85334-161-3

WITH 139 ILLUSTRATIONS

Filmset and printed in Northern Ireland at The Universities Press (Belfast) Ltd.

Preface

The present book is based on and inspired by a series of interpretation-oriented seismic courses given by the author to Shell seismologists and geologists during the period from 1976 to 1981. Familiarity with the various subjects covered is believed to be indispensable for a full comprehension of the significance of reflection information on time profiles, of the nature of reflection interpretation problems and, last but not least, of the present-day limitations of the seismic reflection method. The material comprises a collection of illustrative problems with solutions; these are merged with the text and constitute an integral part of it. References to pertinent books and publications in geophysical journals are cited in chronological order at the end of each chapter. It is hoped that the book will be of interest to geophysical students as well as to those already engaged in reflection interpretation work. The author is indebted to Shell Internationale Petroleum Maatschappij, The Hague, for their permission to publish the lecture material, inclusive of the set of reflection profiles.

A. H. Kleyn

Contents

List of Examples

CHAPTER 1

Introduction

1.1 SEISMIC WAVES

In applied seismology the interpretation of seismic information is almost exclusively based on recordings of a compressional disturbance generated by a seismic source, placed at or below the ground surface or in the water layer.

A compressional wave induces a displacement of rock or water particles which is parallel to the direction of propagation of the wave and is also referred to as a longitudinal wave, pressure wave or P wave.

Typical values for the compressional wave velocity V_p in various types of media are as follows:

Formation	V_p (m/s)
Air	340
Water	1 500
Weathered layer	500–800
Clastics	1 600–4 000
Carbonates	3 000–6 000
Rock salt	4 500
Coal	1 500
Metamorphics	4 000–5 000
Igneous rocks	5 000–6 000

Other types of waves that may be registered during seismic surveys comprise:

1. Shear waves, causing a displacement of subsurface particles in a direction at right angles to the direction of propagation (transverse waves or S waves). A shear wave cannot be sustained by water or by air. Its velocity is usually about half of that of the

compressional wave. The amount of shear wave energy generated by an explosive source is comparatively small, the greater part of the source energy being converted into a P wave disturbance.

2. The surface wave or ground roll, a complex low frequency and high amplitude event which moves along the ground surface with a velocity of the order of 500 to 600 m/s. It has P wave as well as S wave properties, particle displacement components being directed parallel with and at right angles to the direction of propagation.

P waves and S waves travel in volumes of rock and are therefore commonly referred to as body waves, as opposed to surface waves. In a medium of uniform velocity their intensities are inversely proportional to the square of the distance to the source by geometric spreading (spherical divergence). This causes the amplitude of a body wave to decrease inversely with the distance which the wave has covered. For surface waves geometric spreading involves cylindrical divergence, their amplitudes being inversely proportional to the square root of the distance to the source.

1.2 MODE CONVERSION

Mode conversion, or wave-type conversion, is a process whereby part of the P wave energy is converted into S wave energy and vice versa, wherever a P or S wave is obliquely incident on a subsurface discontinuity. Mode conversion also occurs along the ground surface, where part of the emerging body wave energy is transformed into surface waves.

1.3 REFLECTION SHOOTING

The main problem in reflection work is the detection of the reflected primary P wave signal against a background of disturbing events such as ambient noise, surface waves, primary S waves and secondary P and S waves produced by mode conversion.

During data acquisition on land the recording of P waves in preference to S waves is effected by the use of vertical seismometers, or geophones. These, when properly planted, are only sensitive to the vertical component of ground motion, which is relatively large for P waves and small for S waves in view of the near-vertical emergence of reflected waves after their passage through the low velocity weathered zone.

During marine work P waves are commonly detected by pressure sensitive hydrophones, positioned at a depth of about 10 m below the water surface.

The ratio of the root-mean-square amplitude of the reflected primary P wave signal to that of the disturbing events is referred to as the signal to noise ratio. Part of the noise can be considered as systematic, another part as random. Surface waves constitute shot-generated systematic noise, which can be correlated from trace to trace along the seismic record. It can effectively be attenuated by the use of seismometer patterns and by a limited amount of high pass filtering.

Incoherent or random noise includes

1. P and S waves generated by mode conversion during wave propagation.
2. Ambient noise caused by extraneous seismic energy sources such as wind, traffic, rain, general earth unrest etc.

The ambient noise generally differs from shot to shot whereas all shot-generated noise is identical for successive recordings from the same source location.

Methods for attenuating ambient noise and shot-generated random noise comprise the use of seismometer and hydrophone patterns and multi-coverage techniques. The latter involve the stacking of recordings from identical portions of the subsurface (Section 3.1).

1.4 REFRACTION SHOOTING

The interpretation of refraction information is generally based on the travel times of the early arrivals of seismic energy, the ‘first kicks’. These are invariably related to compressional waves since in each formation the propagation velocity of the P wave is larger than that of the S wave.

A refraction process favours mode conversion, the obliquity of refraction ray systems usually being larger than that of reflected rays. Refraction records obtained on land will therefore include a relatively large amount of S wave energy. During refraction work it is only the ambient noise that may interfere with the measurements of first arrival times, however, shot-generated noise and refracted S waves appear on the later parts of refraction records.

BIBLIOGRAPHY

Nettleton, L. L. (1940). *Geophysical Prospecting for Oil*, McGraw-Hill, New York.

DeGolyer, E. (1947). Notes on the early history of applied geophysics in the petroleum industry, *Early Geophysical Papers of the Society of Exploration Geophysicists.*

Weatherby, B. B. (1948). The history and development of seismic prospecting, *Geophysical Case Histories*, **1,** 7–20.

Bullen, K. E. (1949). *Introduction to the Theory of Seismology*, Cambridge University Press, New York.

Schriever, W. (1952). Reflection seismograph prospecting—how it started (history), *Geophysics*, **17,** 936–977.

Dix, C. H. (1952). *Seismic Prospecting for Oil*, Harper & Brothers.

Dobrin, M. B. (1960). *Geophysical Prospecting*, second edition, McGraw-Hill, New York.

O'Brien, P. N. S. (1969). Some experiments concerning the primary seismic pulse, *Geophysical Prospecting*, **17**(4).

O'Brien, P. N. S. and Lucas, A. L. (1971). Velocity dispersion of seismic waves, *Geophysical Prospecting*, **19**(1).

Jolly, R. N. and Mifsud, J. F. (1971). Experimental studies of source generated seismic noise, *Geophysics*, **36**(1).

Scarascia, S., Colombi, B. and Cassinis, R. (1976). Some experiments on transverse waves, *Geophysical Prospecting*, **24**(3).

Sheriff, R. E. (1976). *Encyclopedic Dictionary of Exploration Geophysics*, The Society of Exploration Geophysicists, Tulsa, Oklahoma.

Sweet, G. E. (1978). *The History of Geophysical Prospecting*, Science Press, Los Angeles.

Dohr, G. and Janle, H. (1980). Improvements in the observation of shear waves, *Geophysical Prospecting*, **28**(2).

Lash, C. C. (1980). Shear waves, multiple reflections and converted waves found by a deep vertical wave test (vertical seismic profiling), *Geophysics*, **45**(9).

Wright, C. and Johnson, P. (1982). On the generation of P and S wave energy in crystalline rocks, *Geophysical Prospecting*, **30**(1).

CHAPTER 2

Review of Basic Principles

2.1 ELEMENTS OF SIGNAL THEORY

In terms of signal theory a seismic disturbance can be described as an aperiodic event or transient, also referred to as a wavelet, seismic signal or seismic pulse. On the time scale its duration is roughly of the order of 10 to 50 ms. Correspondingly, for a propagation velocity of 3000 m/s the length of the wavelet would be about 30 to 150 m. When the signal emerging from the source is represented by $S(t)$, the source signature, it may be expressed by $S(t-\tau)$ when recorded after τ s travel time.

The effective duration D of $S(t)$ can be quantified by the relation (Berkhout, 1974)

$$D^2 = \frac{\int_{-\infty}^{+\infty} t^2 S^2(t)\,\mathrm{d}t}{\int_{-\infty}^{+\infty} S^2(t)\,\mathrm{d}t}$$

which takes into consideration how much of the wavelet's total energy occurs during its early part.

According to the Fourier concept a seismic signal can be regarded as the summation of an infinite number of cosine functions or Fourier components, each being characterized by frequency, amplitude and phase. The amplitude spectrum of the signals gives the relationship between the amplitudes of Fourier components and corresponding frequencies, whereas the phase spectrum relates phase angles to frequencies.

For aperiodic events spectra are continuous. An amplitude spectrum is then a density spectrum, the amplitude density being expressed in amplitude units per unit of frequency, such as volt seconds. The actual amplitude of a Fourier component for the frequency f is therefore given by the infinitesimal quantity $A(f)\,df$.

A single zero phase Fourier component of amplitude density $A(f)$ is represented by $A(f)\,df \cos 2\pi ft$, which modifies to $A(f)\,df \cos 2\pi f(t-\tau)$ after a time shift τ in the direction of the positive time axis. The phase angle is then $-2\pi f\tau$ radians, which is referred to as a phase lag. Similarly, an advance of τ seconds to earlier time produces a phase lead of $2\pi f\tau$ radians.

For a given time function phase and amplitude spectra can be computed by means of a Fourier transform, an inverse Fourier transform generating a time function corresponding to given amplitude and phase spectra.

Amplitude spectra of seismic wavelets are band limited, frequencies ranging from about 6 to 200 Hz (hertz, or cycles/s). This implies that the seismic signal has no zero frequency component and that therefore its average amplitude is zero. During propagation part of the wavelet's energy is dissipated by absorption and scatter. These are frequency dependent processes, the higher frequency components of the signal spectrum being attenuated more rapidly than the lower frequency ones. This causes a progressive shift of the more significant part of the amplitude spectrum toward the lower frequencies and a corresponding broadening of the wavelet. The vertical resolution of the reflection method will therefore decrease with increasing depth of investigation.

For a zero phase wavelet phase angles are zero for all frequencies, so that zero phase signals are symmetrical about their time origin. They belong to the general category of physically non-realizable signals, having non-zero amplitudes at negative times. During seismic processing recorded reflection signals may be transformed into their zero phase equivalents. This special processing technique plays an important role in the stratigraphic interpretation of reflection information (Section 5.4).

Other significant properties of zero phase signals are related to amplitude and effective duration. With regard to amplitude it can be shown that of all signals with identical amplitude spectra the zero phase signal has the largest maximum amplitude. Or, representing the zero phase equivalent of $S(t)$ by $S_0(t)$, we have

$$S_0(0) \geqslant S(t) \qquad \text{for any value of } t$$

Similarly, it can be demonstrated that of all signals with identical amplitude spectra the zero phase signal has the shortest effective duration. Consequently, signal to noise ratio and vertical resolution will be enhanced by zero phasing.

A special type of zero phase signal is the delta pulse or 'spike', having a flat amplitude spectrum with unit amplitude density. In the time domain it is given by

$$\delta(t)=0 \qquad \text{for } t\neq 0$$
$$\delta(t)=\infty \qquad \text{for } t=0$$

$$\text{with} \quad \int_{-\infty}^{+\infty} \delta(t)\,\mathrm{d}t=1$$

The delta pulse can be considered as an ideal, non-attainable, seismic signal, providing optimum vertical resolution; its band limited version in the form of a zero phase signal is sometimes referred to as a pseudo delta pulse.

Most explosive sources emit minimum phase signals, which are the closest approximations to zero phase signals in the domain of physically realizable signals. Important properties related to minimum phase signals are as follows:

1. When $S_{\mathrm{m}}(t)$ and $S(t)$ are minimum phase and non-minimum phase signals with identical amplitude spectra, it obtains that

   $$\int_0^t S_{\mathrm{m}}^2(t)\,\mathrm{d}t \geqslant \int_0^t S^2(t)\,\mathrm{d}t \qquad \text{for each value of } t$$

   which means that the energy of the minimum phase wavelet is optimally concentrated toward its onset (minimum energy delay property).
2. There exists a unique relationship between phase and amplitude spectra of minimum phase signals.
3. Of all physically realizable signals with identical amplitude spectra the minimum phase signal has the shortest effective duration.

2.2 PRINCIPLE OF FERMAT

Fermat's principle states that the propagation time of a seismic pulse between two points A and B is equal to the travel time measured along a path of minimum or maximum time connecting A and B. Such paths are called raypaths, wave paths or Fermat paths. Raypaths are usually minimum time paths, the maximum time condition being restricted to special synclinal features (Section 7.5). For any source–detector pair there are in general a multitude of minimum time connections, each being related to a specific seismic event such as a refraction, a shallow reflection, a deep reflection etc.

2.3 SNELL'S LAW

Snell's law describes the change of direction of a Fermat path where it strikes an interface separating media of different velocities. With

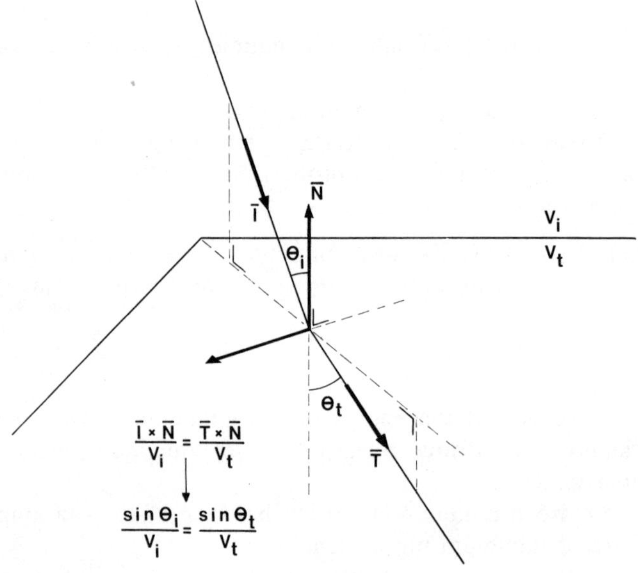

FIG. 2.1. Snell's law.

reference to Fig. 2.1, we have in consequence of Fermat's principle

1. $$\frac{\sin\theta_i}{\sin\theta_t}=\frac{V_i}{V_t}$$

where θ_i and θ_t are the ray's angles of incidence and transmission respectively.

2. Incident and transmitted ray segments are situated in a common plane oriented at right angles to the plane of the velocity interface.

The relationship between Snell's law and the Fermat principle is illustrated in Fig. 2.2.

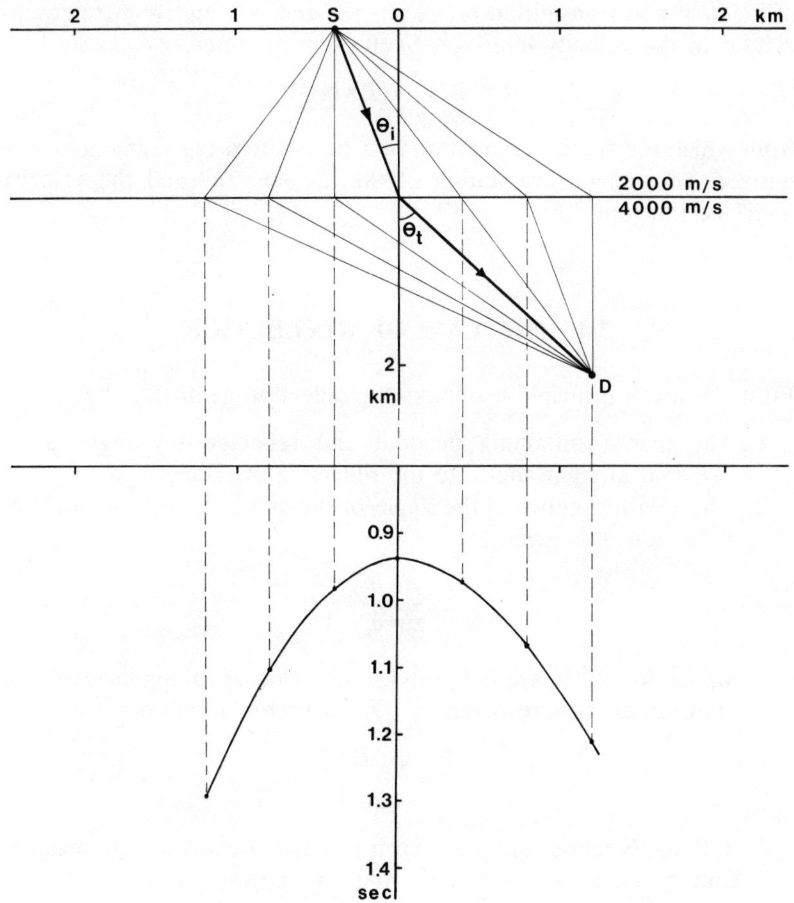

FIG. 2.2. Illustration of Fermat's principle.

For the special case that the angle of transmission is 90° the transmitted ray segment will be referred to as a refracted path. The angle of incidence is then the critical angle given by the relation

$$\sin \theta_c = V_i/V_t \qquad \text{for } V_i < V_t$$

From Fig. 2.1 a straightforward three-dimensional ray tracing algorithm may be evolved by expressing Snell's law as an equation of vector cross-products. When **I** and **T** are unit vectors in the directions of incident and transmitted ray segments, and **N** a unit vector perpendicular to the velocity interface, Snell's law becomes

$$(\mathbf{I} \times \mathbf{N})/V_i = (\mathbf{T} \times \mathbf{N})/V_t$$

from which the spatial orientation of a transmitted ray segment can be derived for a given orientation of the incident ray and the velocity interface (Section 3.5).

2.4 THE LAW OF REFLECTION

From Fermat's principle it obtains for reflection geometry that

1. The plane containing incident and reflected ray segments is oriented at right angles to the reflecting interface.
2. The relation between the angle of incidence θ_i and the angle of reflection θ_r is given by

$$\frac{\sin \theta_i}{\sin \theta_r} = \frac{V_i}{V_r}$$

where V_i and V_r are the propagation velocities along incident and transmitted rays respectively. Or, in vector notation,

$$\frac{\mathbf{I} \times \mathbf{N}}{V_i} = \frac{\mathbf{R} \times \mathbf{N}}{V_r}$$

I, **R** and **N** representing unit vectors in the directions of the incident and reflected ray segments and the normal to the reflecting interface.

The above relations are general formulations of the reflection law. They include the case of the PP and SS reflections for which the angle of incidence equals the angle of reflection, and that of the PS and SP reflections associated with mode conversion, as illustrated in Fig. 2.3.

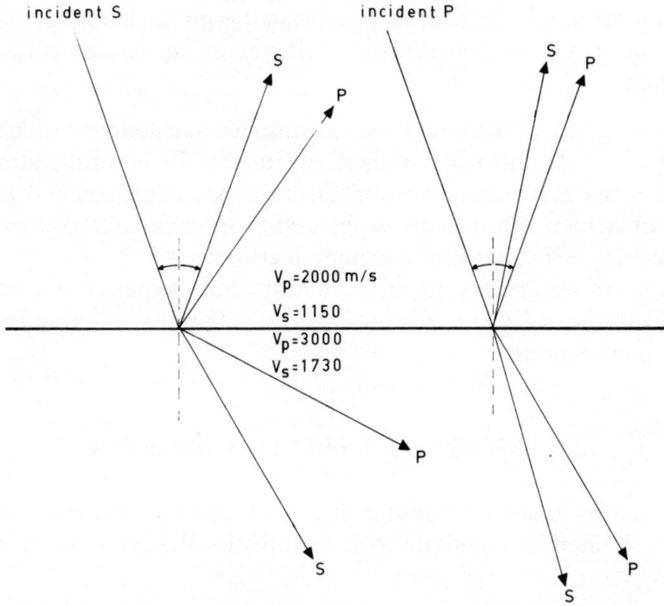

FIG. 2.3. Wave-type conversion during reflection process. Example of relationships between angles of incidence, reflection and transmission.

For most practical applications, however, V_i and V_r correspond to the compressional wave velocity in the medium of incident and reflected rays.

2.5 DIFFRACTION

Diffraction of seismic energy is a process whereby part of incident seismic energy is diffused by acoustic discontinuities in the form of points or linear elements, sometimes referred to as point reflectors and line reflectors.

Geological conditions which may produce diffraction events comprise

1. Truncation of reflectors by fault planes.
2. Truncation of reflectors by unconformities.
3. Truncation of reflectors by salt, igneous or shale diapirs.
4. Onlap of reflectors against unconformities.

5. Lateral lithologic contrasts associated with a change of depositional conditions, such as sand lenses or silt stones encased in shales.

The above diffraction sources are examples of line sources, which may be horizontal or inclined, straight or curved. In nature, diffraction point sources are rather exceptional. They are approximated by the apexes of slender salt or igneous intrusions or, in the offshore area, by shipwrecks, passing ships and drilling platforms.

In terms of seismic ray geometry, a diffraction raypath is a least time path connecting seismic source and receiver via the relevant diffraction line or point source.

2.6 KIRCHOFF–FRESNEL CONCEPTS

Fig. 2.4 shows an acoustic source at point S and a pressure detector at point P, situated in a medium with compressional wave velocity V and

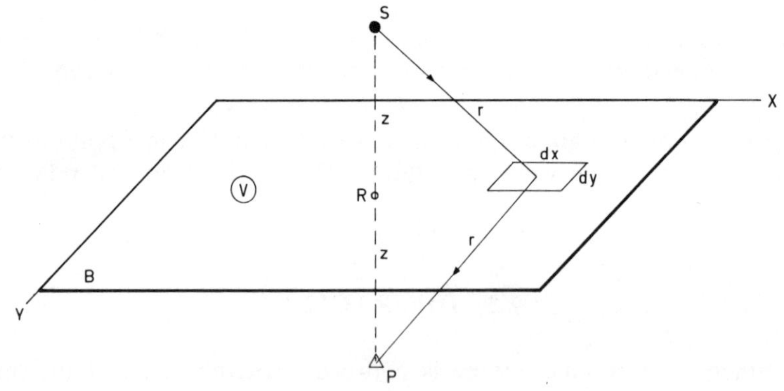

FIG. 2.4. Geometry of Kirchoff integral for symmetrical case.

separated by an acoustically transparent plane B. Representing the source signature by $p_S(t)$, we have for the pressure field

$$p(r, t) = \frac{p_S(t - r/V)}{r}$$

the factor $1/r$ expressing the effect of spherical divergence.

The pressure recording at P can be related to hypothetical recordings in plane B by means of the Kirchoff integral expression

$$p_P(t) = \frac{z}{2\pi} \int_{-\infty}^{+\infty} \int_{-\infty}^{+\infty} \left(\frac{1}{r^3} + \frac{1}{Vr^2} \frac{\partial}{\partial t} \right) p_S\left(t - \frac{2r}{V} \right) \mathrm{d}x\, \mathrm{d}y$$

When plane B extends to infinity, the solution of the Kirchoff integral becomes

$$p_P(t) = \frac{p_S(t - 2z/V)}{2z}$$

as could be expected. For a restricted extent of plane B, however, its solution is less simple and will include a diffraction wave field.

Applying the above to a zero-offset reflection recording, point S could represent a seismic source at the earth's surface with coincident detector and plane B a reflecting interface. Then, the Kirchoff summation does not basically change when regarding point P as the mirrored detector position. After introducing a diffraction edge by truncating plane B, the reflector becomes the equivalent of a transparent half plane, bounded by a non-transparent complement. The response of such a reflection–diffraction model at a sequence of coincident source–detector locations along a straight line traverse comprises the familiar hyperbolic alignment of diffracted signals which culminates at the crossing of a reflection traverse and a diffraction line source. The forward branch of the diffraction curve will have the same polarity as the reflection event, its backward branch exhibiting a phase inversion. These phenomena cannot always be clearly observed in practice, however. The same is true for a postulated 50% reduction of reflection amplitude at the diffraction edge.

Kirchoff's wave acoustics indicates that a reflection ray system carries only geometrical significance and that the actual reflection process should be explained in terms of contributions to the build up of the reflection signal by an infinite number of elements of the reflecting interface. These can be grouped in Fresnel zones whose geometry is illustrated in Fig. 2.5 for the case of normal incidence at a horizontal reflector. Let point S be a source emitting a monochromatic signal with wavelength λ. The central Fresnel zone is defined as the circular area surrounding the reflection point R for which $2(r-z) \leq \frac{1}{2}\lambda$. The remaining region of the reflector can be subdivided in consecutive Fresnel rings with R as centre and where contributions to the Kirchoff integral

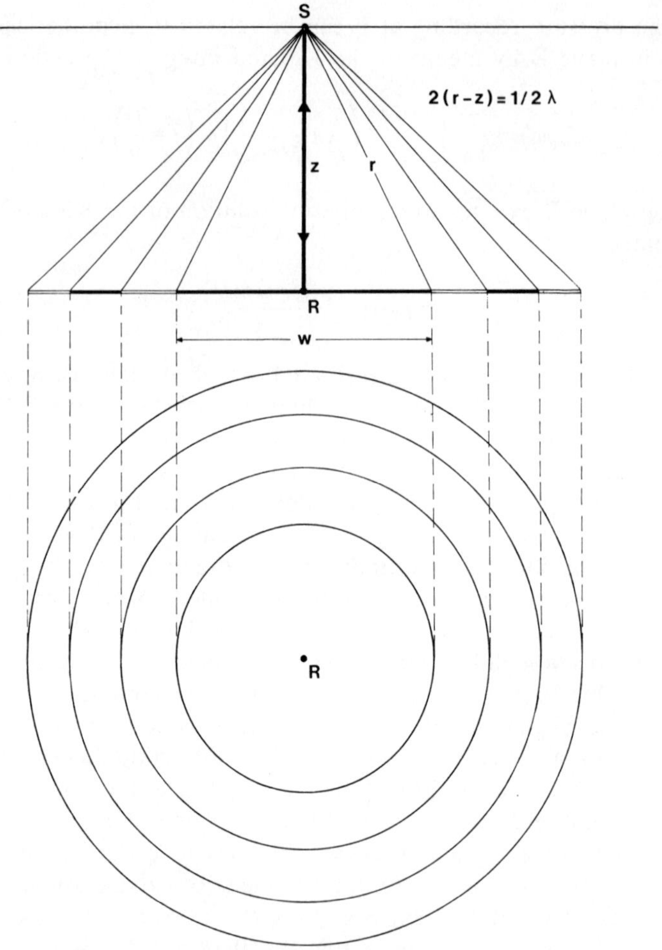

FIG. 2.5. Fresnel zones for normal incidence reflection.

are alternately in-phase and 180° out-of-phase with that of the central Fresnel zone. The condition for in-phase contribution can then be written as

$$2n\pi < 2\pi(2r/\lambda - 2z/\lambda) < (2n+1)\pi$$

$$\text{or} \qquad n\lambda < 2(r-z) < (n+\tfrac{1}{2})\lambda \qquad \text{for } n = 1, 2, 3, \ldots$$

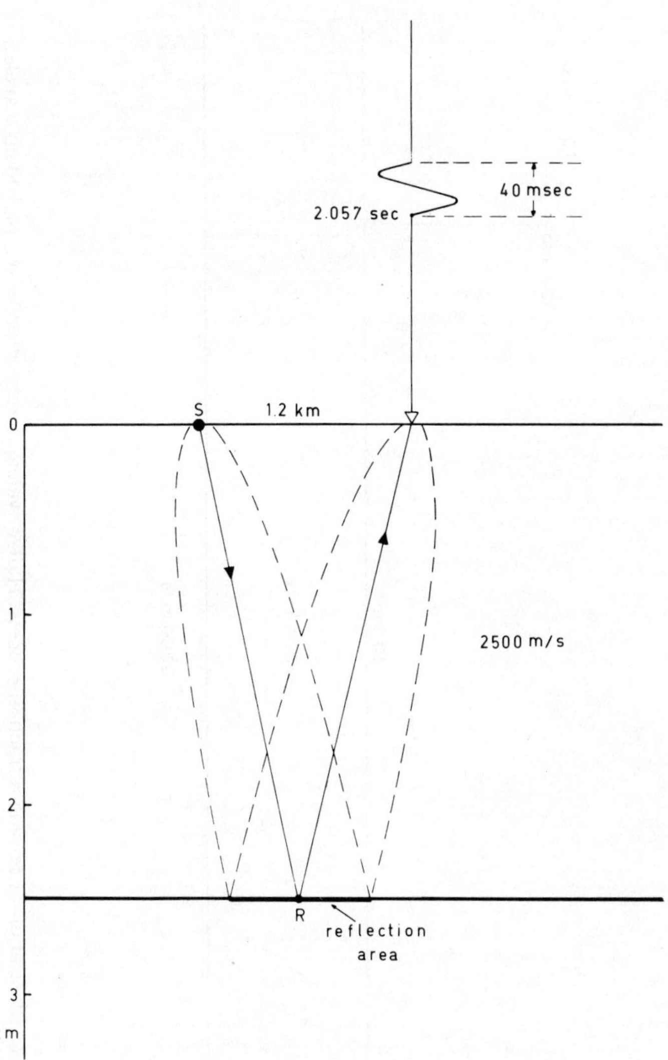

FIG. 2.6. Illustration of seismic reflection process. Half-wavelength reflection beam for 40 ms wavelet.

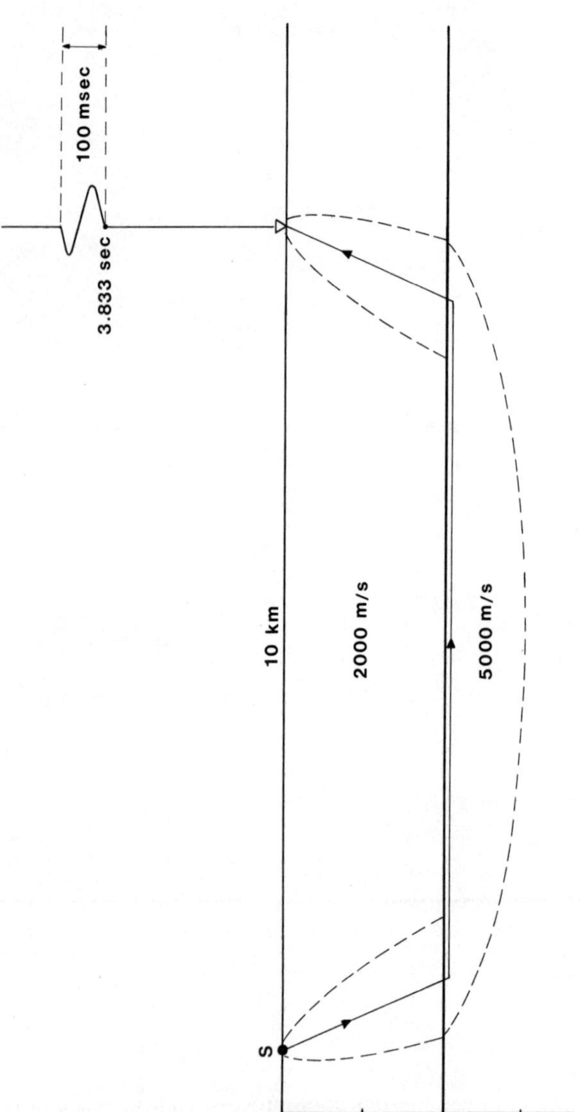

FIG. 2.7. Illustration of seismic refraction process. Half-wavelength refraction beam for 100 ms wavelet.

Similarly, for rings with out-of-phase contributions we have

$$(m-\tfrac{1}{2})\lambda < 2(r-z) < m\lambda \qquad \text{for } m=1,2,3,\ldots$$

The Kirchoff integral shows that the magnitudes of the Fresnel ring contributions diminish in the lateral direction and that the more significant contributions originate from the central Fresnel zone. From Fig. 2.5, the width of this zone is given by

$$w=(2z\lambda+\tfrac{1}{4}\lambda^2)^{\frac{1}{2}}$$

$$\text{or} \qquad w\simeq(2z\lambda)^{\frac{1}{2}} \qquad \text{for } z\gg\lambda$$

which indicates that the larger the reflector depth and the predominant wavelength in the signal spectrum the lower the lateral resolution. Signal migration (Chapter 8) effectively lowers the recording plane to deeper levels, narrowing the Fresnel zones and therefore enhancing lateral resolution, which is often referred to as 'focusing'.

Figs 2.6 and 2.7 show approximate rock volumes involved in the transfer of seismic energy during reflection and refraction processes. Fig. 2.7 refers to the condition that for the frequency range normally encountered during seismic exploration refracted wavelets can only be observed where the refracting layer has a sufficient magnitude. It may also be intuitively obvious that a high speed refractor having the thickness of a razor blade could not transport a significant amount of low frequency refracted energy, though Fermat's principle would correctly postulate the refraction times.

The above considerations also explain frequency selective scattering of seismic energy by subsurface inhomogeneities. Such a scattering process can be expected to be more severe for the higher frequency Fourier components, associated with relatively narrow energy beams. Similarly, the scatter of the sun's rays in the atmosphere causes a blue sky.

2.7 WAVEFRONTS: HUYGENS' PRINCIPLE

In seismic geometry wavefronts are defined as surfaces of equal travel time, associated with a single seismic source and with a specific type of event. We may thus distinguish between reflection and refraction wavefronts.

From the Fermat principle there exists in isotropic media an orthogonal relationship between wavefront surfaces and raypaths. Accordingly, when the velocity distribution is represented by $V(x, y, z)$, a wavefront for time $t + \mathrm{d}t$ is the envelope tangent to a family of elementary spheres with radii $V(x, y, z)\,\mathrm{d}t$ and centres located at the wavefront for time t. This describes the Huygens principle as applied to forward extrapolation of a wavefront surface. The inverse procedure of deriving wavefronts in the direction of earlier times is called backward extrapolation, a standard refraction interpretation technique.

In wave theory both the forward and the backward extrapolation of wave fields can be quantified by means of appropriate versions of the Kirchoff integral. The term 'downward extrapolation', associated with the migration of reflection signals, is equivalent to backward extrapolation. The Kirchoff integral for backward extrapolation is the basis of the Kirchoff migration system, to be discussed in Section 8.2.

2.8 PRINCIPLE OF RECIPROCITY

In ray acoustics the reciprocity principle refers to the condition that for a specific event the travel time from a source at point A to a detector at point B is the same as that from a source at B to a detector at A. This property is an obvious implication of the Fermat principle.

2.9 TIME–DISTANCE CURVES

A plot of the travel time of a seismic event versus source–detector offset along a straight line traverse is called a time–distance curve or t–x plot. At any point along a t–x profile the time dip is the absolute value of the derivative $\mathrm{d}t/\mathrm{d}x$, its reciprocal being referred to as the apparent velocity V_{a}. The direction of the time dip corresponds to the direction of increase of travel time.

2.10 NEAR-SURFACE SEISMIC GEOMETRY

With reference to Fig. 2.8, let us consider a narrow bundle of parallel rays emerging at a point P, and a system of associated wavefronts. The angle of emergence $i(0)$ of a ray is defined as the angle it makes with

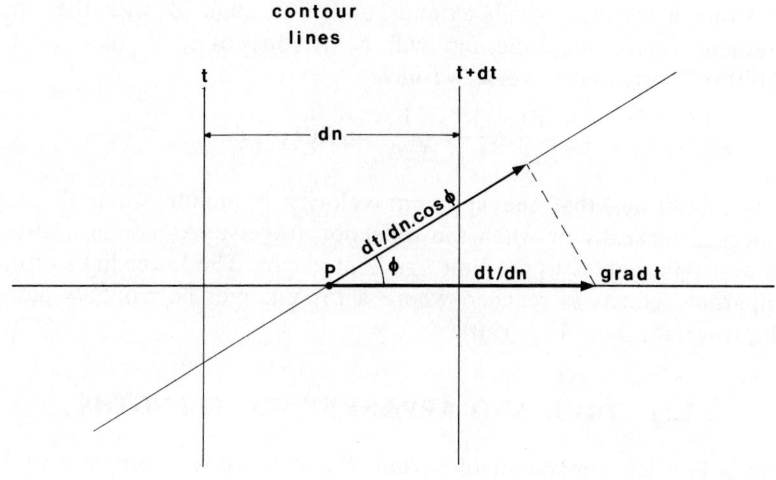

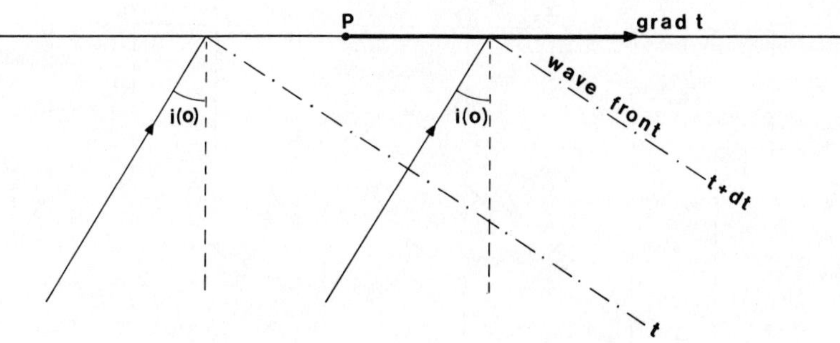

FIG. 2.8. Near-surface seismic geometry.

the vertical. Wavefront outcrops in the vicinity of P are lines of equal travel time, or time contours, at right angles to the emerging rays. The time dip along a profile through point P extending in the direction of the time gradient vector is dt/dn, dt being an elementary contour interval and dn the corresponding contour spacing. From Fig. 2.8,

$$\frac{\mathrm{d}t}{\mathrm{d}n}=\frac{1}{V(0)}\sin i(0)$$

where $V(0)$ is the velocity just below the ground surface.

Along a traverse which extends under an angle ϕ with the time gradient vector the time dip will be $(\mathrm{d}t/\mathrm{d}n)\cos\phi$, so that for an arbitrarily oriented traverse we have

$$\frac{\mathrm{d}t}{\mathrm{d}x}=\frac{1}{V_{\mathrm{a}}}=\frac{1}{V(0)}\sin i(0)\cos\phi$$

which indicates that the apparent velocity is infinite when the ray emerges vertically or when the reflection traverse extends in a direction at right angles to the time gradient vector. The lower limit of the apparent velocity is reached where a ray emerges horizontally along the traverse; then $V_{\mathrm{a}}=V(0)$.

2.11 TRUE AND APPARENT WAVELENGTHS

For a Fourier component of period T and a propagation velocity V, the true wavelength is VT. Similarly, the apparent wavelength λ_{a} is given by $\lambda_{\mathrm{a}}=V_{\mathrm{a}}T$.

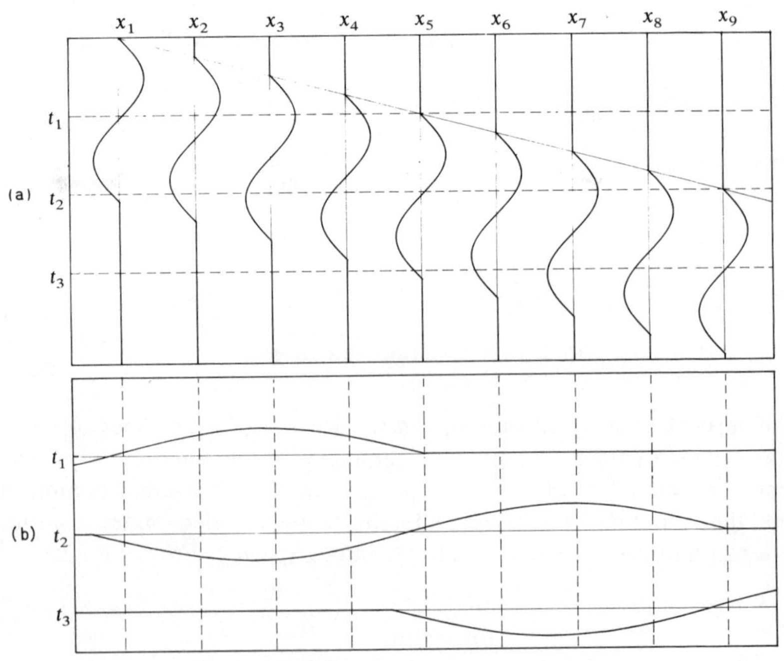

FIG. 2.9. t–x profiles for one-cycle cosine wave.

The geophysical meaning of apparent wavelength and apparent velocity may follow from a study of t–x profiles. Fig. 2.9(a) is a display of recordings of a one-cycle cosine wave in the form of time traces at surface positions $x_1, x_2, x_3, \ldots$, Fig. 2.9(b) showing recordings of the same event in the form of x traces for the time instants $t_1, t_2, t_3, \ldots$. The x traces in Fig. 2.9(b) illustrate the waveform along the ground surface as seen by an observer at the given time instants. He would then measure the apparent values for wavelength and wave speed and establish the relation $\lambda_a = V_a T$.

2.12 *k-f* DIAGRAMS AND DETECTOR ARRAYS

The reciprocal of the apparent wavelength is the wave number k or spatial frequency expressed in cycles per unit distance. Wave number and frequency are comparable quantities in the Fourier analysis of waveforms in space and time domains. They are related to each other by $f = V_a k$, in consequence of $\lambda_a = V_a T$. For a uniform value of V_a (k, f) pairs plotted in a k–f reference system will be distributed along a straight line passing through its origin with an inclination of V_a m/s. Posted amplitude density values in the k–f plane define a k–f spectrum or 'k–f diagram'.

k–f diagrams can be derived by means of a two-dimensional Fourier transform of recordings in the x–t domain. When a low velocity surface wave interferes with reflections of high apparent velocities a k–f spectrum will show concentrations of significant spectral contributions in zones encompassing different apparent velocity lines. From Fig. 2.10, this may lead to a separation of wave number spectra where frequency spectra overlap so that the surface wave may be suppressed by k-domain filtering.

A detector array is an example of a k-domain filter. Its elements straddle a station location and are connected to a single recording channel. For a simple linear array consisting of N detectors spaced at a distance d, the normalized wave number response R is given by

$$R = \frac{\sin \pi N \,\mathrm{d}k}{N \sin \pi \,\mathrm{d}k}$$

The array response curve included in Fig. 2.10 indicates that a detector array acts as a low pass k-filter, the boundary between pass and suppression bands corresponding to the wave number $1/Nd$, which is the reciprocal of the array length.

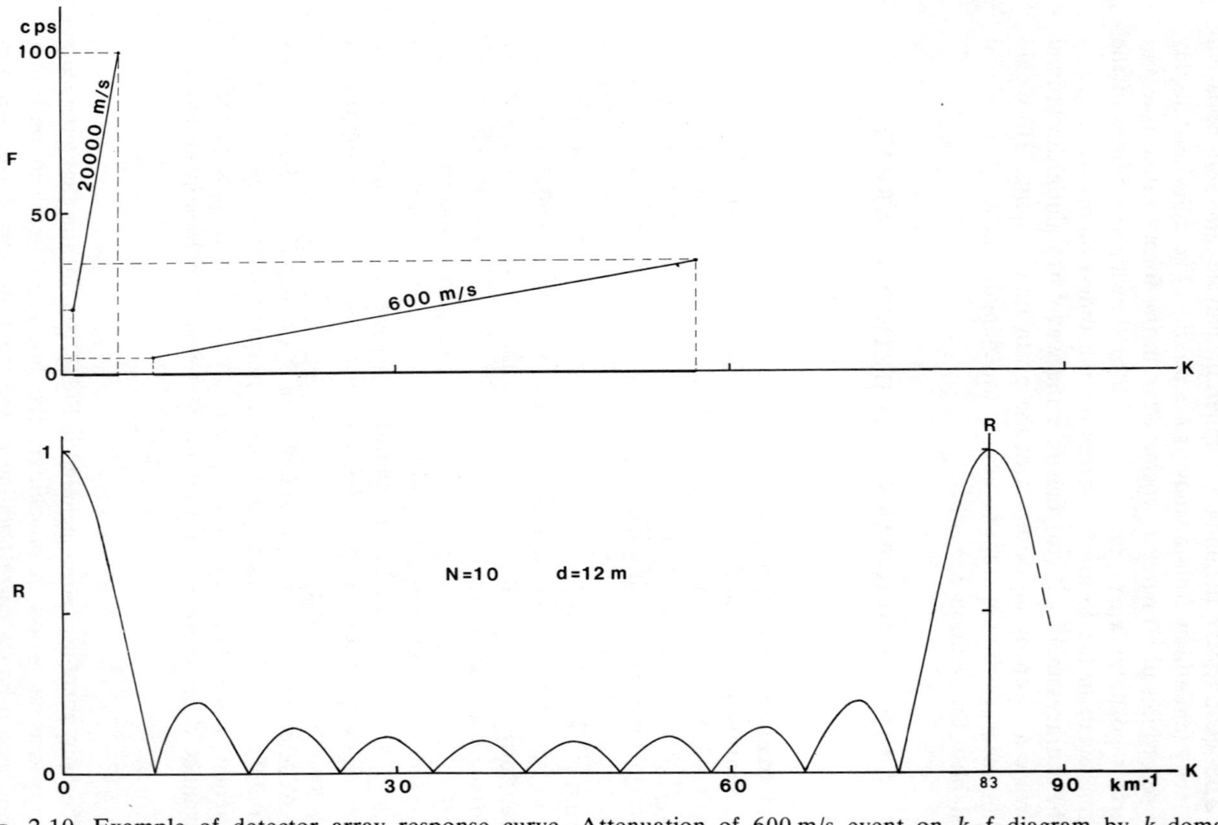

FIG. 2.10. Example of detector array response curve. Attenuation of 600 m/s event on k–f diagram by k-domain filtering.

In terms of Fourier theory the wave number response R is the k-spectrum of an x-domain filter consisting of a sequence of delta functions placed at detector positions. The general theory includes negative k values. The response curve is then periodic in the k interval extending from plus infinity to minus infinity, with period $1/d$. It is further of interest to note that by alternating the polarity of an even number of pattern elements, the response formula changes into

$$R = \frac{\sin \pi N\, dk}{N \cos \pi\, dk}$$

which represents a bandpass filter with $R = 1$ for $k = 1/2d$.

Example 2.1 *Detector array*
A reflection event emerges under an angle of 30° in the vertical plane of a seismic traverse. The frequency band extends from 15 to 150 Hz. The near-surface velocity is 2000 m/s. Determine

1. The wave number bandwidth.
2. The maximum length of a linear detector array which will pass the 60 Hz frequency component with an attenuation of not more than 6 dB when the detector spacing is 5 m.

Example 2.1 *Solution*
From $V_a = V(0)/\sin i(0)$, $V_a = 3000$ m/s.
From $k = f/V_a$, $k_{min} = 0{\cdot}005$ and $k_{max} = 0{\cdot}050$ cycles/metre (cpm).
For $f = 60$ Hz, $k = 0{\cdot}020$ cpm.
Using 7 elements, $R = 0{\cdot}37$; in decibels: $20 \log 0{\cdot}37 = -8{\cdot}5$ dB.
Using 6 elements, $R = 0{\cdot}51$; in decibels: $20 \log 0{\cdot}51 = -5{\cdot}8$ dB.
Required number of elements is six; array length 30 m.

2.13 SAMPLING AND ALIASING

The array response curve in Fig. 2.10 is the equivalent of the amplitude spectrum of a sampled box of unit height and having a width which is equal to the array length Nd. By decreasing the sampling interval d the period of its wave number spectrum will increase. When d tends to zero the spectrum becomes aperiodic and identical to the amplitude spectrum of the continuous box.

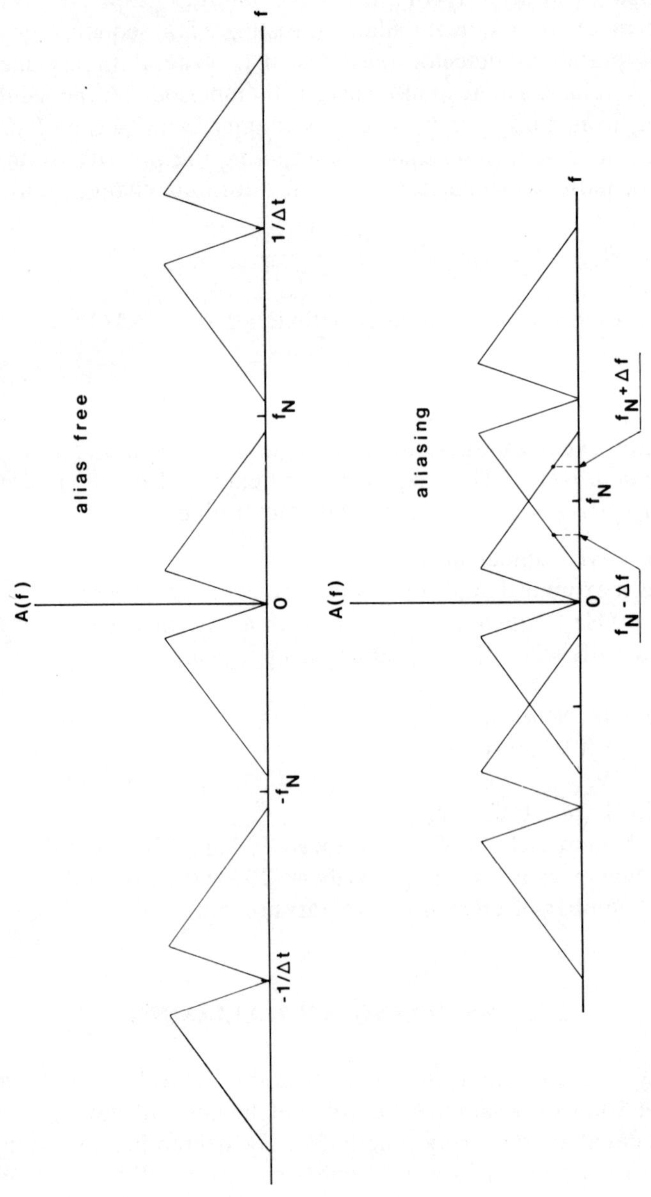

FIG. 2.11. Amplitude spectra of sampled time signal.

Correspondingly, it can be shown that the frequency spectrum of the sampled version of a seismic signal consists of periodic repetitions of the spectrum of the continuous signal, as illustrated in Fig. 2.11. For a sampling interval of Δt seconds the spectrum's period is $1/\Delta t$ Hz. Half that value is the Nyquist frequency f_N. Correct sampling requires that f_N is larger than the highest significant signal frequency. Otherwise the periodic elements of the spectrum of the sampled signal will overlap, destroying the fidelity of the sampling procedure. A Fourier component of frequency $f_N + \Delta f$ of the continuous signal will then appear as an aliased component of frequency $f_N - \Delta f$ after digital-to-analogue

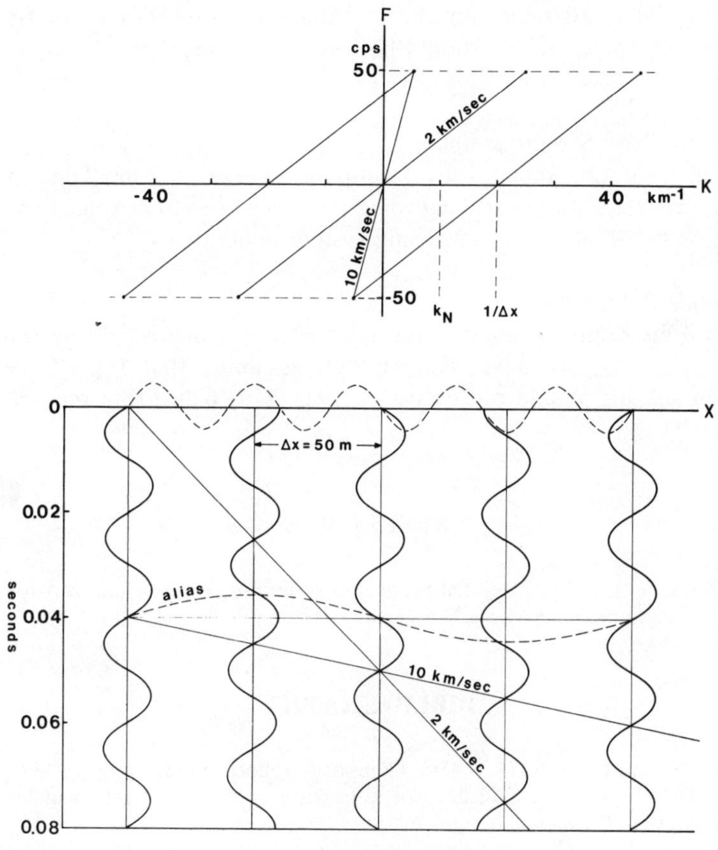

FIG. 2.12. Aliasing in k–f and x–t domains.

conversion. To avoid aliasing, spectral contributions above the Nyquist frequency will be removed during data acquisition by a low pass filter, referred to as an anti-alias filter.

The Nyquist sampling criterion must be fulfilled for time functions as well as for space functions. When the trace interval on reflection records is Δx, the highest significant wave number should not exceed the Nyquist wave number $1/(2\Delta x)$. A detector array will therefore be an anti-alias filter when its length is twice the station spacing.

Fig. 2.12 illustrates aliasing in k–f and x–t domains for a 50 Hz cosine wave having a velocity of 2000 m/s. Its wave number is 0·025 cpm. Employing a trace spacing of 50 m the Nyquist wave number is 0·010 cpm. This causes an alias of 0·005 cpm, corresponding to velocities of 10 000 m/s in the k–f and x–t planes. For an alias-free system the trace spacing should be less than 20 m.

Example 2.2 *Spatial sampling*

A high resolution survey uses a sampling interval of 1 ms. Determine the appropriate station spacing when the lowest apparent velocity to be handled by an alias-free recording system is 6000 m/s.

Example 2.2 *Solution*

For a 1 ms sampling interval the Nyquist frequency is 500 Hz. From $k_{max} = f_{max}/V_{min}$ we have $k_{max} = 0{\cdot}083$, assuming that $f_{max} = f_N$. The station spacing should not exceed $1/(2k_{max})$, or 6 m. From the above, $\Delta x/\Delta t = V_{min}$.

REFERENCE

Berkhout, A. J. (1974). Related properties of minimum-phase and zero-phase time functions, *Geophysical Prospecting*, **22**(4).

BIBLIOGRAPHY

Ricker, N. and Lynn, R. D. (1950). Composite reflections, *Geophysics*, **15**(1).

Krey, T. (1952). The significance of diffraction in the investigation of faults, *Geophysics*, **17**, 843.

Trorey, A. W. (1970). A simple theory for seismic diffractions, *Geophysics*, **35**(5).

Berkhout, A. J. (1973). On the minimum-length property of one-sided signals, *Geophysics,* **38**(4).

Berkhout, A. J. (1974). Related properties of minimum-phase and zero-phase time functions, *Geophysical Prospecting,* **22**(4).

White, R. E. and O'Brien, P. N. S. (1974). Estimation of the primary seismic pulse, *Geophysical Prospecting,* **22**(4).

Hamming, R. W. (1977). *Digital Filters,* Prentice-Hall Inc., Englewood Cliffs, New Jersey.

Anstey, N. A. (1977). *Seismic Interpretation: the Physical Aspects,* International Human Resources Development Corporation, Boston, USA.

Payton, C. E. (1977). Seismostratigraphy—applications to hydrocarbon exploration, section 1, *Memoir 26,* The American Association of Petroleum Geologists.

Levin, F. K. (1978). The reflection, refraction and diffraction of waves in media with an elliptical velocity dependence, *Geophysics,* **43**(3).

Sheriff, R. E. (1980). Nomogram for Fresnel-zone calculation, *Geophysics,* **45**(5).

Meissner, R. and Hegazy, M. A. (1981). The ratio of the PP- to SS-reflection coefficient as a possible future method to estimate oil and gas reserves, *Geophysical Prospecting,* **29**(4).

Anstey, N. A. (1971). *Signal Characteristics and Instrument Specifications,* volume 1 of *Seismic Prospecting Instruments,* Gebrüder Borntraeger, Berlin–Stuttgart.

CHAPTER 3

Geometrical and Analytical Backgrounds

3.1 BASIC REFLECTION GEOMETRY

An ideal reflection time section is the noise-free seismic response of the subsurface to a data acquisition system consisting of sources and coincident detectors, uniformly distributed along a straight line traverse. This definition is requisite to an understanding of dynamic correction procedures, which convert recordings from spreads of detectors into zero-offset traces. The latter are the principal components of an unmigrated reflection time section and can be considered as recordings of reflected wavelets which have been travelling along raypaths at normal incidence on reflecting interfaces, as illustrated in Fig. 3.1. Reflection sections are therefore referred to as 'normal incidence' sections. An important aspect of the normal incidence reflection geometry in Fig. 3.1 is the horizontal displacement of a reflection point with reference to a corresponding data point along the recording traverse. This is the migration shift, which may in general have components parallel to the section or at right angles to it, depending on the orientation of the seismic traverse with respect to the direction of subsurface dip.

General definition of dynamic corrections

The use of zero-offset data acquisition systems is restricted to shallow investigations such as measuring water depths with an echo sounder and sub-sea-bottom profiling. Deep penetration reflection work is normally carried out by shooting into spreads of recording stations, spread lengths varying from about 1200 to 2500 m, depending on the depths of the objectives.

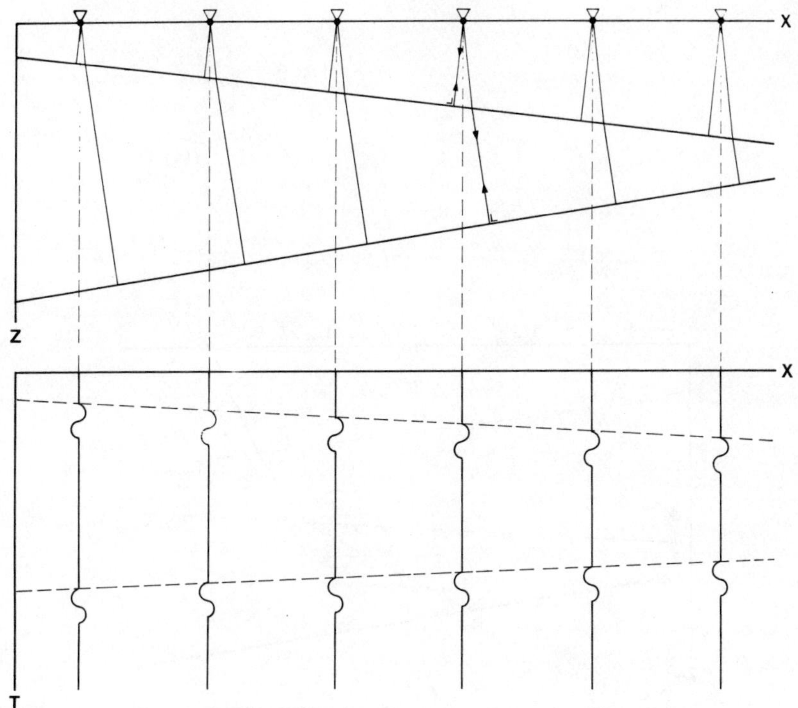

FIG. 3.1. Normal incidence reflection geometry.

Referring to Fig. 3.2, the dynamic shift or moveout is for each shot–station pair defined as the difference between a recorded reflection time and the corresponding normal incidence reflection time at the midpoint of the shot–station interval.

Horizontal reflector

For a horizontal reflector underlying a constant velocity overburden the reflection t–x curve is a branch of the hyperbola

$$t_x^2 = t_0^2 + \frac{x^2}{V^2}$$

where $t_0 = 2z/V$, z being the reflector depth and V the velocity. For large shooting distances the slope of the reflection t–x curve tends to the limit $1/V$, reflected rays emerging near-horizontally. This implies

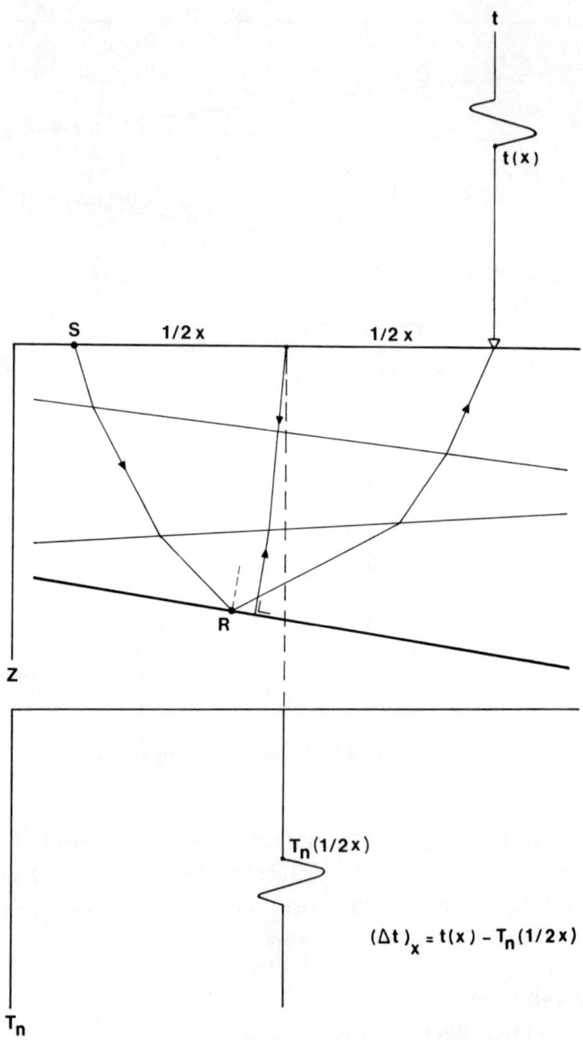

FIG. 3.2. Definition dynamic correction.

that the asymptotes of the reflection hyperbola are $t = \pm x/V$, which is also the t–x relation for the direct wave, travelling in the horizontal direction with velocity V.

For the horizontal reflector model the normal incidence reflection time is equal to t_0 at any point along the traverse. Therefore, if the moveout for the shooting distance x is $(\Delta t)_x$, it obtains that

$$(\Delta t)_x = t_x - t_0 = t_x - (t_x^2 - x^2/V^2)^{\frac{1}{2}}$$

From the general relationship between the slope of a t–x curve and the angle of emergence (Section 2.10) the curvature of a sequence of reflection hyperbola will decrease with increase of recording time. This explains the expression 'dynamic corrections' as opposed to 'static corrections' which are related to surface topography and near-surface weathering effects.

The parabolic approximation of the reflection hyperbola follows from

$$t_x = t_0(1 + x^2/V^2 t_0^2)^{\frac{1}{2}} \approx t_0(1 + x^2/2V^2 t_0^2) = t_0 + x^2/2V^2 t_0$$

which implies that the dynamic correction is nearly proportional to the square of the shooting distance.

We may further conclude that for the present basic case the overburden velocity can be determined by means of a t^2–x^2 plot of the field data. This will yield a straight line with slope $1/V^2$.

Inclined reflector

Fig. 3.3 shows the reflection geometry for a non-horizontal reflector and a uniform velocity. The reflection t–x graph is a hyperbola whose axis of symmetry passes through the mirror image of the shot point about the reflecting interface. On a reflection field record the vertex of the reflection hyperbola and its asymptotes will then be displaced in the updip direction with reference to the position of the shot point, whereas the updip branch of the reflection hyperbola terminates at the t–x graph of the direct wave.

For the computation of moveout values it will be convenient to express the t–x relation of the reflection event in terms of the normal incidence times at midpoint positions. Denoting these by $T_n(\frac{1}{2}x)$, we have from Fig. 3.4

$$t_x^2 = T_n^2(\tfrac{1}{2}x) + x^2/V_s^2$$

where $V_s = V/\cos\delta$, δ being the dip angle of the reflecting interface.

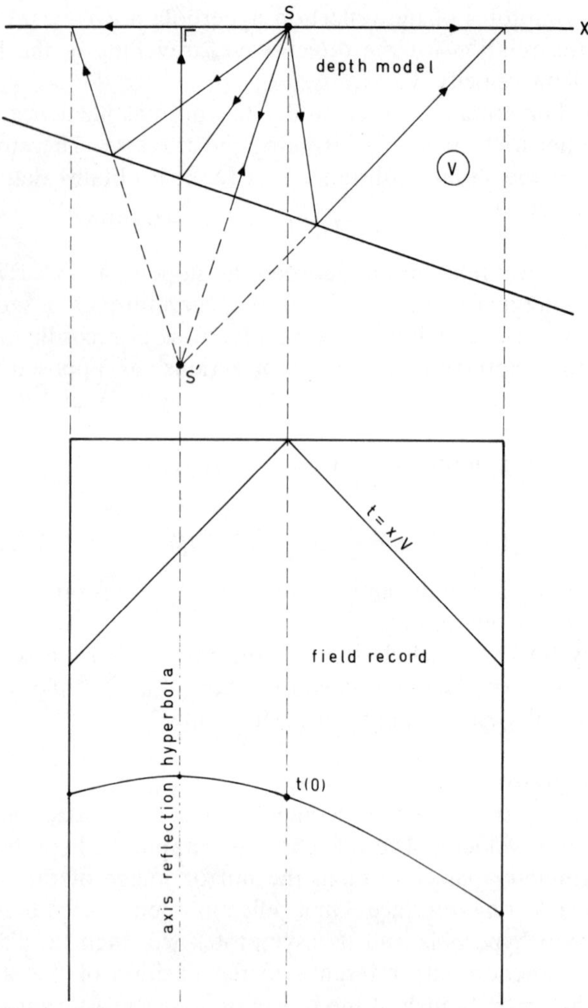

FIG. 3.3. Reflection t–x graph for inclined reflector.

Then, by analogy with the case of the horizontal reflector, the dynamic correction is given by

$$(\Delta t)_x = t_x - (t_x^2 - x^2/V_s^2)^{\frac{1}{2}}$$

For any position of the spread with respect to that of the shot point the interval between midpoints associated with both ends of the spread

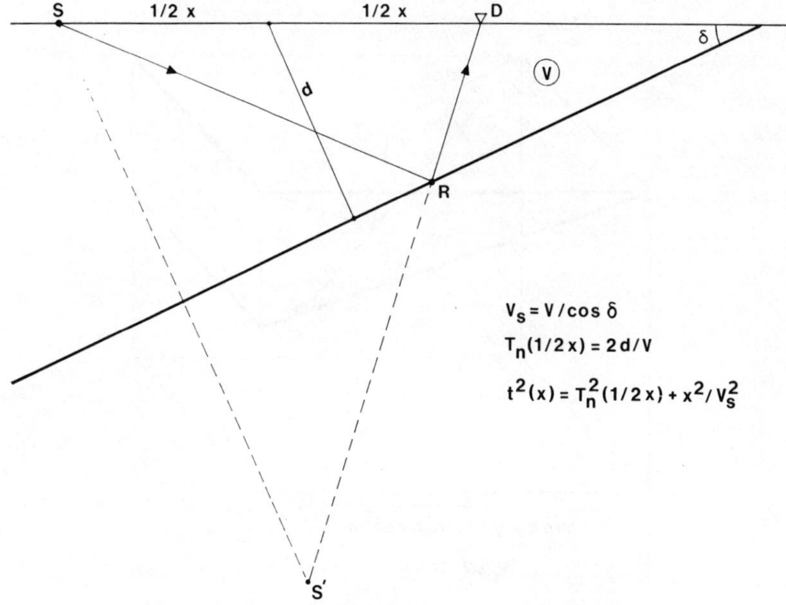

FIG. 3.4. Relationship between normal incidence reflection time $T_n(\frac{1}{2}x)$ at shot–detector midpoint and observed reflection time $t(x)$.

equals half the spread length. This is called the subsurface coverage. Continuous subsurface coverage requires successive shifts of at least half a spread length of the data gathering system. Illustrative examples of reflection sections obtained by this older single coverage survey method are shown on profiles A and Q. Section Q is an early reflection recording revealing evidence of a gas–water contact in a gas-bearing anticline. Section A shows the type of early 1960s profiles in the Netherlands which led to the discovery of the Slochteren gas field in the Groningen area.

The above definition of V_s is only true when the reflection profile extends in the direction of subsurface dip. For an arbitrarily oriented traverse the dip angle should be measured in the plane of the reflection ray system (Section 2.4), so that from Fig. 3.5 we have in general

$$V_s = V(1 - \sin^2 \delta \cos^2 \omega)^{-\frac{1}{2}}$$

where ω is the angle between the directions of reflection traverse and reflector dip.

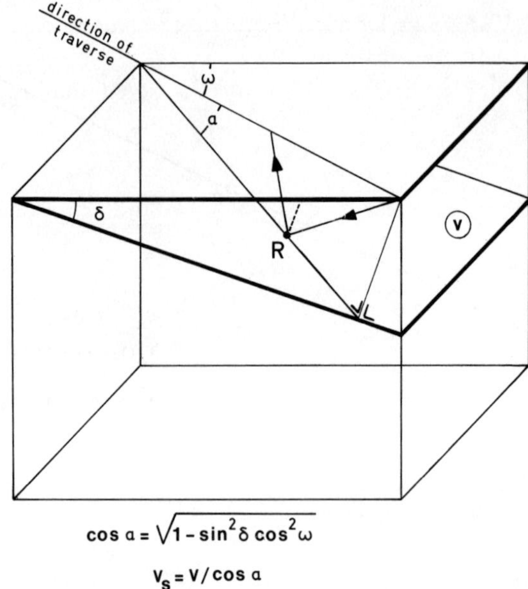

FIG. 3.5. Dependency of V_s on reflector dip and direction of reflection traverse.

The minimum value for V_s is the overburden velocity V ($\delta = 0$). For the trivial case that $\delta = 90°$ and $\omega = 0$, V_s becomes infinite. Then $t_x = T_n(\frac{1}{2}x)$, as also follows from reflection ray geometry.

Example 3.1 *Basic reflection geometry*

A reflection traverse is parallel to the strike of a reflecting interface. Reflector dip δ degrees; vertical depth of reflector below traverse z m; velocity V m/s. Derive the equation of the reflection t–x curve.

Example 3.1 *Solution*

$$t_x^2 = t_0^2 + x^2/V^2 \qquad \text{with } t_0 = \frac{2z\cos\delta}{V}$$

considering that the plane containing the reflection raypaths is oriented at right angles to the reflecting interface.

Example 3.2 *Moveout velocity, dynamic corrections*
For a centre-spread recording along a dip traverse the maximum shooting distance is 1200 m. It is further given that $t_{1200} = 1{\cdot}061$ s, $t_{-1200} = 1{\cdot}239$ s, $t_0 = 0{\cdot}985$ s, overburden velocity = 2000 m/s and reflector dip = 10°.

1. Compute V_s and plot the dynamically corrected reflection section for the given data set.
2. Transform graphically the three-trace reflection section into a migrated depth section and verify that the structural dip is 10°.
3. Using the results of (2), determine by ray tracing the t–x coordinates of the vertex of the reflection hyperbola.

Example 3.2 *Solution*
The solution of this example is shown in Fig. 3.6.

Multi-coverage data acquisition
For a data gathering system consisting of source–receiver combinations arranged along a straight line traverse and with a common midpoint it obtains with reference to Fig. 3.7 that

$$t_x^2 = t_0^2 + x^2/V_s^2$$

where x is the source–receiver offset and t_0 the normal incidence reflection time at the common midpoint ($x = 0$), replacing the previously used symbol $T_n(\frac{1}{2}x)$.

In the above expression t_0 is a constant and a t–x plot will yield a one-sided hyperbola, ending at right angles to the time axis at $x = 0$. The reflection hyperbola becomes two-sided and symmetrical about the time axis when including negative x values, corresponding to reciprocal source–receiver positions. Fig. 3.7 shows the relevant signal display, commonly referred to as a point sort or common midpoint gather.

The principal objective of common midpoint data acquisition is the increase of the signal to noise and primary to multiple ratios (Section 3.9) by stacking the dynamically corrected traces of the point sort. In that respect it should be noted that by varying the shooting geometry, shot-generated noise caused by wave-type conversion will differ from trace to trace. The summation of traces will therefore attenuate both

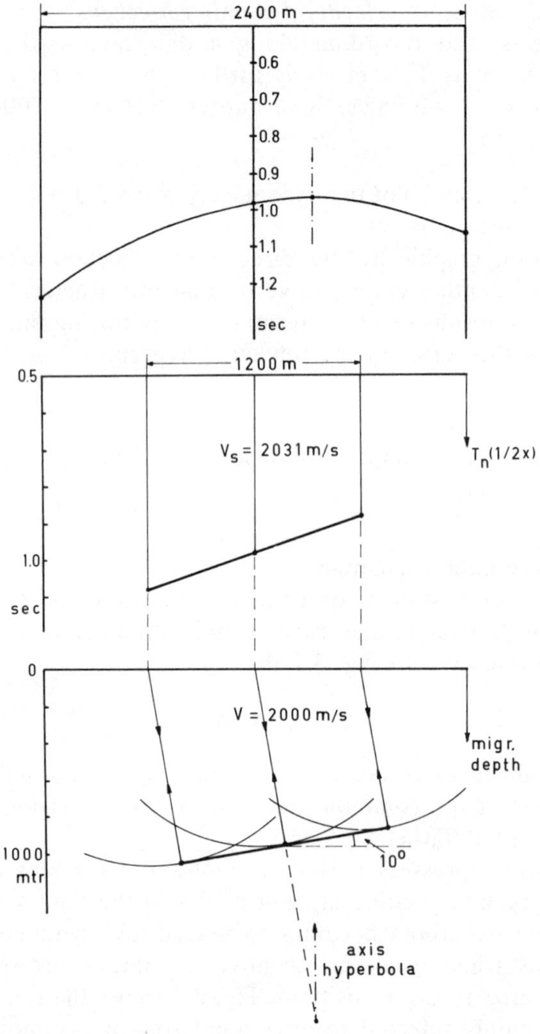

FIG. 3.6. Example 3.2 solution.

the shot-generated noise and the ambient noise by destructive interference, whereas the coherent primary reflection signals will be amplified by in-phase addition.

The effectiveness of the stacking process depends on the number of traces added, assuming perfect static and dynamic corrections. An n-fold stack has an r.m.s. noise level which is $\sqrt{n}$ times that of a single

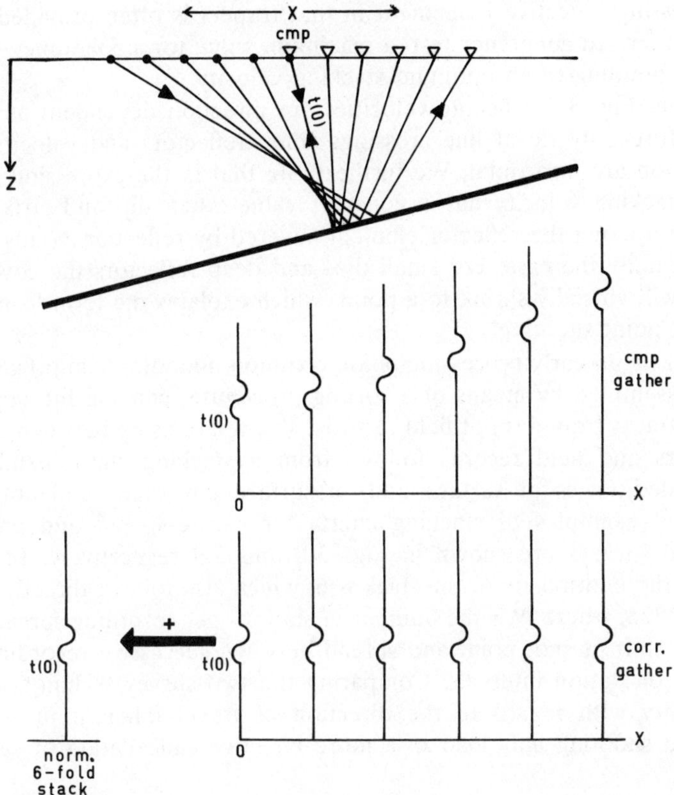

FIG. 3.7. Principle of common midpoint reflection geometry and signal stack.

trace, whereas reflection amplitudes are multiplied by n. The signal to noise ratio improvement factor is therefore $\sqrt{n}$. A 16-fold stack will thus have a signal to noise ratio which is twice that of a 4-fold stack, and so on.

The quantity V_s can now be referred to as the stacking velocity and the dynamic correction written in the form

$$(\Delta t)_x = (t_0^2 + x^2/V_s^2)^{\frac{1}{2}} - t_0$$

The appropriate value for V_s can be determined by a trial and error procedure, testing a range of trial velocities until signals on dynamically corrected traces align at a common t_0 level. The above expression for $(\Delta t)_x$ indicates that too large a value for V_s will result in undercorrection and too small a value in overcorrection. During digital processing objective judgement in that respect is often provided by a computerized coherency test, a maximum value for a coherency factor corresponding to an optimum stacking velocity.

From Fig. 3.5, stacking velocities are direction dependent and will therefore only tie at line crossings when reflectors and velocity distribution are horizontal. We further note that in the expression for t_x^2 the stacking velocity has a constant value when dip and strike are uniform along the reflector element covered by reflection points. This is normally the case. For small dips and deep reflectors the coverage area will virtually shrink to a point, which explains the term 'common depth point shooting'.

During an early processing phase common midpoint (c.m.p.) gathers are assembled by means of a sorting procedure, copying the appropriate traces from a set of field records. The relationship between c.m.p. gathers and field records follows from a stacking chart, exhibiting exploded views of surface and subsurface coverage configurations. Typical examples of stacking charts for centre-spread and end-of-spread surveys are shown in Figs 3.8 and 3.9 respectively. In both cases the multiplicity of the stack is 6, which also follows directly from $M = N/2S$, where N is the number of stations per recording spread and S the shift of shot point and spread between successive recordings in terms of station intervals. Comparing the two survey techniques, the diversity with regard to the direction of travel inherent in centre-spread shooting may lead to a more effective cancellation of surface noise.

End-of-spread shooting using a streamer containing 48 or 96 hydrophone groups is the standard offshore data acquisition technique. Land surveys employ end-of-spread as well as centre-spread arrangements with station intervals of 25 or 50 m. For all reflection field systems the maximum shot–station offset will normally not exceed about 2500 m.

Signal stretch

From basic reflection geometry the moveout of a reflection signal is determined by velocity, shooting distance and travel time. On account of the composite character of reflection events and by noise interference the arrival times of the individual reflection wavelets can never be measured exactly, however. Therefore, the practical realization of a dynamic correction procedure entails a dynamic shift of each sample of the reflection trace without reference to the actual signal onsets. This causes the early part of the reflection signal to move over a slightly

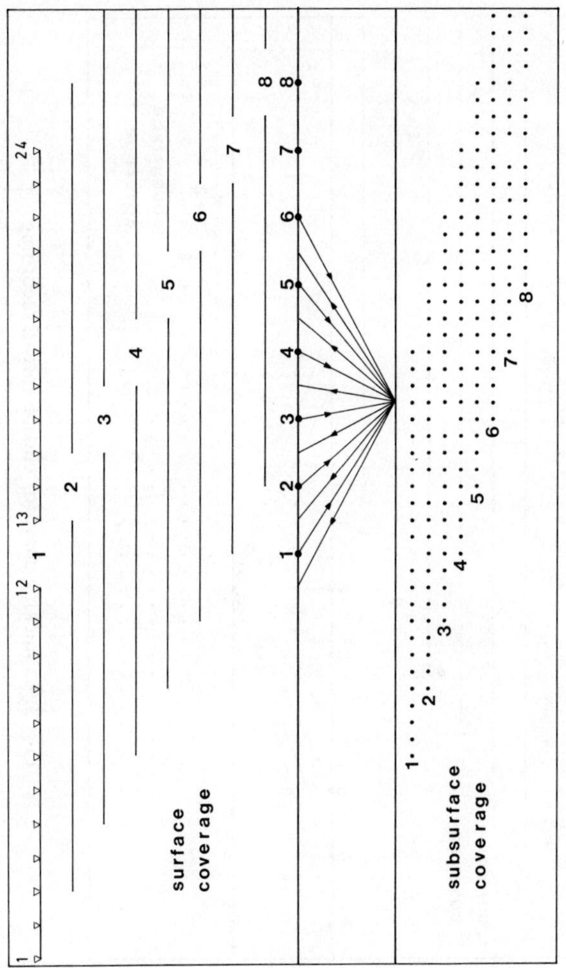

FIG. 3.8. Stacking chart for centre-spread recording.

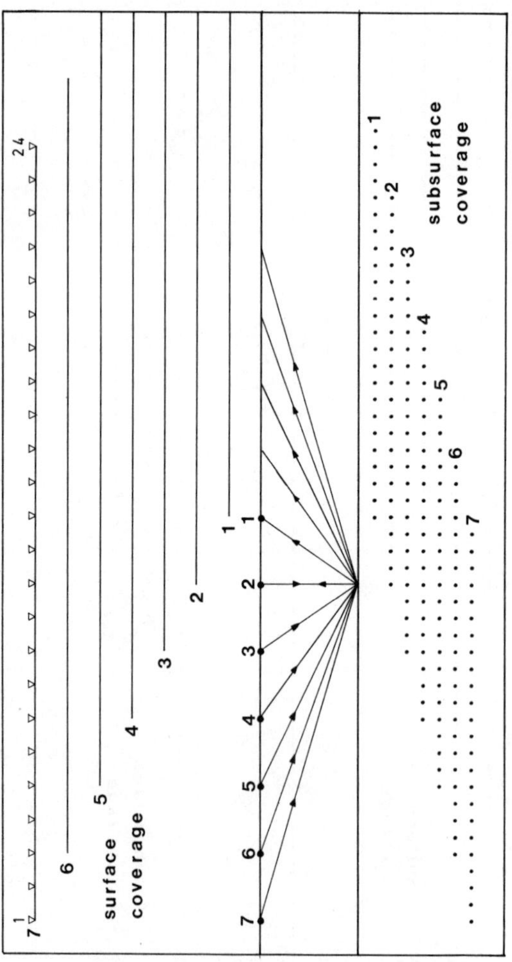

FIG. 3.9. Stacking chart for end-of-spread recording.

larger time interval than its tail, resulting in signal stretch. For given values of x and t_x the stretch factor S is defined as the ratio of the extension of a small trace element to its duration prior to the moveout correction, or

$$S = \frac{dt_0 - dt_x}{dt_x} = \frac{dt_0}{dt_x} - 1$$

From $t_0^2 = t_x^2 - x^2/V_s^2$ we have

$$\frac{dt_0}{dt_x} = \frac{t_x}{t_0}$$

so that the stretch factor in terms of velocity, zero-offset time and shooting distance becomes

$$S = \frac{1}{V_s t_0}(t_0^2 V_s^2 + x^2)^{\frac{1}{2}} - 1$$

which defines lines of equal stretch in the t–x plane. These will be straight for a uniform V_s and curved for a more general stacking velocity distribution.

In view of the adverse effect of signal stretch on stacking response stretch factors must be below a tolerable limit. The value of the limiting stretch factor depends on the nature of the processing problem. Stretch factors should be relatively small, for instance, when dealing with high frequency information. On the other hand, larger stretch factors may be acceptable when putting emphasis on multiple suppression (Section 3.9). Typical values for the limiting stretch factor are of the order of 20%. It is standard practice to erase after the dynamic correction procedure all trace intervals with stretch factors which exceed the specified limit. This is called blanking or muting, causing a reduction of the multiplicity of the stack in the shallower parts of the reflection section. The inverse slope of a line of equal stretch corresponding to the limiting stretch factor is referred to as the blanking velocity.

Example 3.3 *Signal stretch*

For a horizontal reflector underlying a 2500 m/s velocity medium the zero-offset reflection time is 1·100 s. Determine the signal stretch and stretch factor after a dynamic correction at $x = 3000$ m when the signal's duration is 0·032 s.

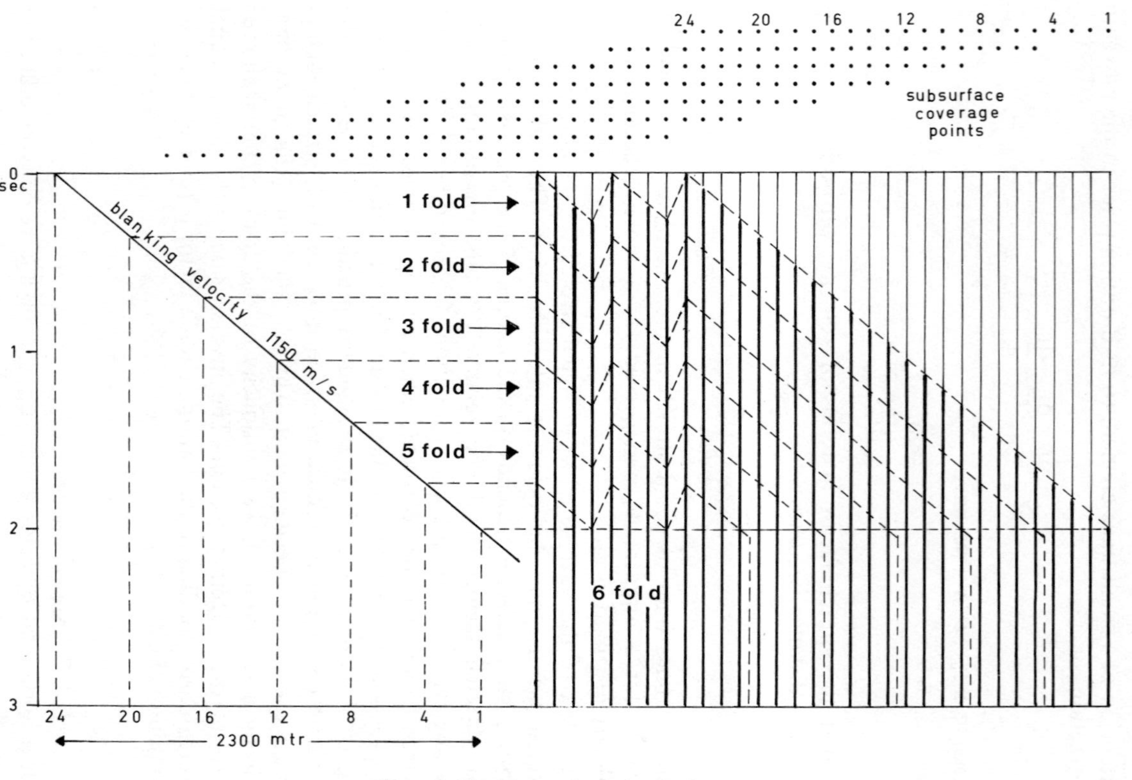

FIG. 3.10. Example 3.4 solution.

Example 3.3 *Solution*
From $t_x^2 = t_0^2 + x^2/V^2$ the signal onset is at 1·628 s. The dynamic shift of the signal's tail at 1·660 s is 0·513 s and that of its onset 0·528 s. The signal stretch is therefore 15 ms and the average stretch factor is 47%.

Example 3.4 *Multiplicity of stack after blanking*
Principal parameters of a multi-coverage end-of-spread data acquisition system are as follows:

24 station recording spread
Station interval 100 m
Shot point interval 200 m (shot point trails spread)
Maximum shot–station offset 2300 m
Outline zones of equal multiplicity for the initial subsurface coverage interval of 1550 m when the blanking velocity is 1150 m/s.

Example 3.4 *Solution*
From $M = N/2S$ the theoretical multiplicity is 6. Fig. 3.10 shows the variation of actual multiplicities with depth after blanking.

Plane wave response simulation
An alternative approach to the processing of reflection field data is based on plane wave reflection geometry. Seismic energy emerging from a plane wave source along the ground surface will initially descend in a vertical direction. From Fig. 3.11, showing plane wave reflection geometry for a uniform velocity, a plane wave reflection profile could be assembled without resorting to a dynamic correction procedure. The practical realization of this concept involves a simulation of the response to a plane wave source from the data of a conventional multi-coverage data acquisition system. The method has only a restricted application but is interesting from a fundamental point of view. When we have a line source along a dip traverse, the reflection response to the cylindrical wave field would be identical to that to a plane wave from an areal source. For a sampled version of the line source in the form of a sequence of closely spaced point sources the simulation principle is based on the following considerations

1. The downgoing wave is the envelope of a set of elementary Huygens wavefronts emitted by the point sources.

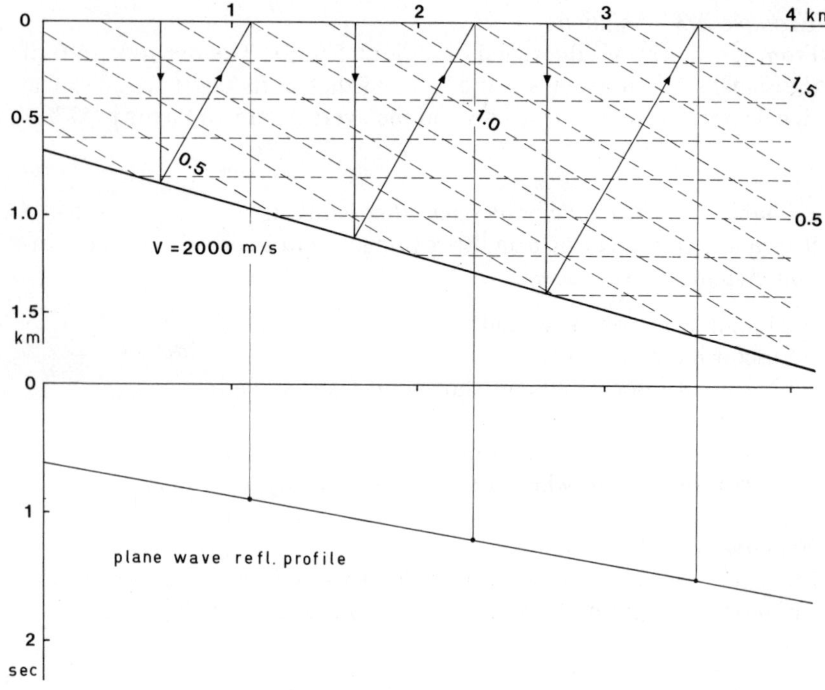

FIG. 3.11. Plane wave reflection geometry.

2. The recording of the reflected cylindrical wave at a central receiver will be identical to the stack of recordings from the individual point sources.
3. In view of seismic reciprocity we may replace the point sources by detectors and the central receiver by a point source. The result in (2) can therefore also be obtained by directly stacking the recordings from a conventional centre-spread shot.

For a line source of sufficient lateral extent the reflection time of the stack is that of the vertex of the reflection hyperbola, irrespective of reflector dip. The method can be extended to handle steep dip cases by retarding the traces recorded at successive detectors by prescribed time intervals, simulating reflection response to an inclined wavefront. Fig. 3.12 illustrates plane wave response simulation for end-of-spread data acquisition.

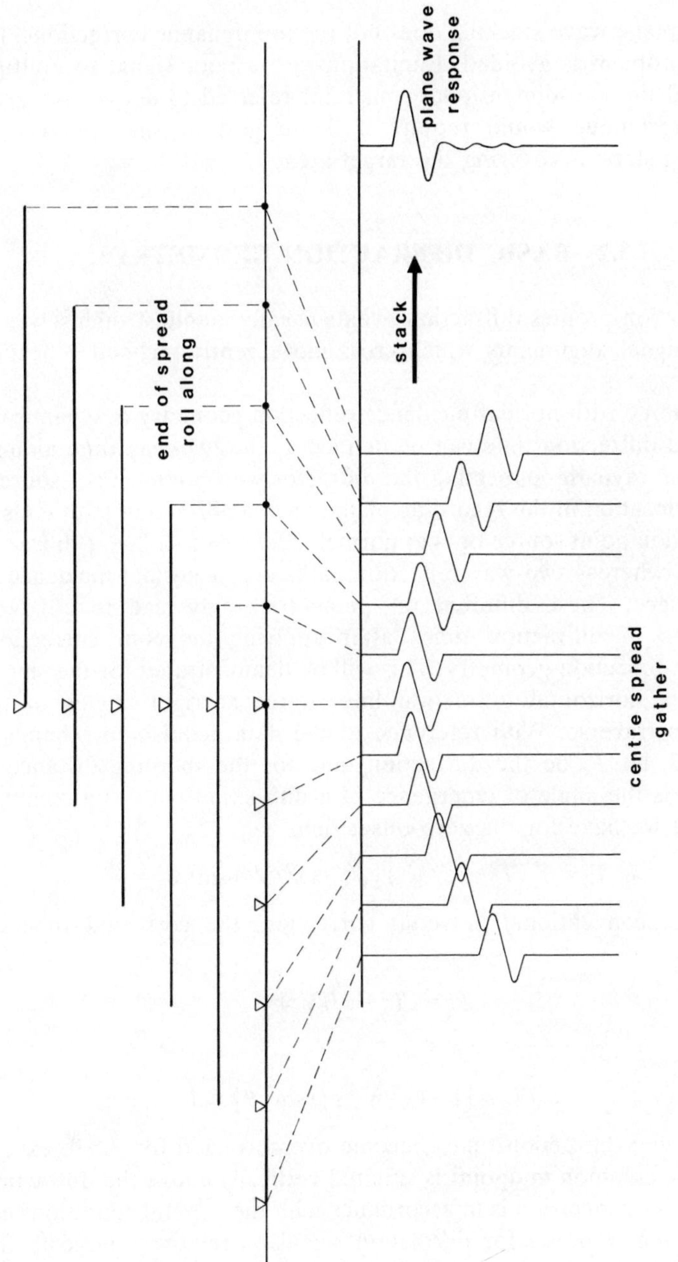

FIG. 3.12. Plane wave response simulation from end-of-spread reflection data.

Since plane wave stacking does not require dynamic corrections, the stretch problem is avoided. Limitations are a poor signal to multiple ratio and the two-dimensional constraint referred to above. An areal survey technique would require a dense grid of shot points and recording stations covering the target area.

3.2 BASIC DIFFRACTION GEOMETRY

On reflection profiles diffraction events usually manifest themselves as curved signal alignments which cross more gently inclined reflection sequences.

In analogy with normal incidence reflection geometry a dynamically corrected diffraction time can be defined as the two-way time along a diffraction raypath connecting the diffractor with a coincident source–receiver location in the recording plane. Such a diffraction path ends at a diffraction point source or is at normal incidence to a linear diffractor element, whereas two-way reflection paths are at normal incidence to an interface. These different ray geometries may lead to an over-correction of diffraction times after applying moveout corrections based on reflection geometry. This will be demonstrated for the simple case of a horizontal diffraction line source at right angles to the reflection traverse. With reference to the data acquisition scheme in Fig. 3.13, let T_x be the diffraction time for the shooting distance x. When θ is the angle of emergence of a diffraction path at a common midpoint we have for the zero-offset time

$$T_0 = T_x(T_x^2 - x^2/V^2)^{\frac{1}{2}}[T_x^2 - (x^2/V^2)\sin^2\theta]^{-\frac{1}{2}}$$

After the conventional moveout correction, the corrected time T_0' becomes

$$T_0' = (T_x^2 - x^2/V^2)^{\frac{1}{2}}$$

so that

$$T_0'/T_0 = [1 - (x^2/V^2 T_x^2)\sin^2\theta]^{\frac{1}{2}} \leq 1$$

showing that diffraction times become overcorrected for $x > 0$, except where the common midpoint is situated vertically above the diffraction source. This conclusion is in accordance with the general condition that the stacking response for diffraction signals is relatively large in the region of the culmination of a diffraction curve and diminishes along its flanks.

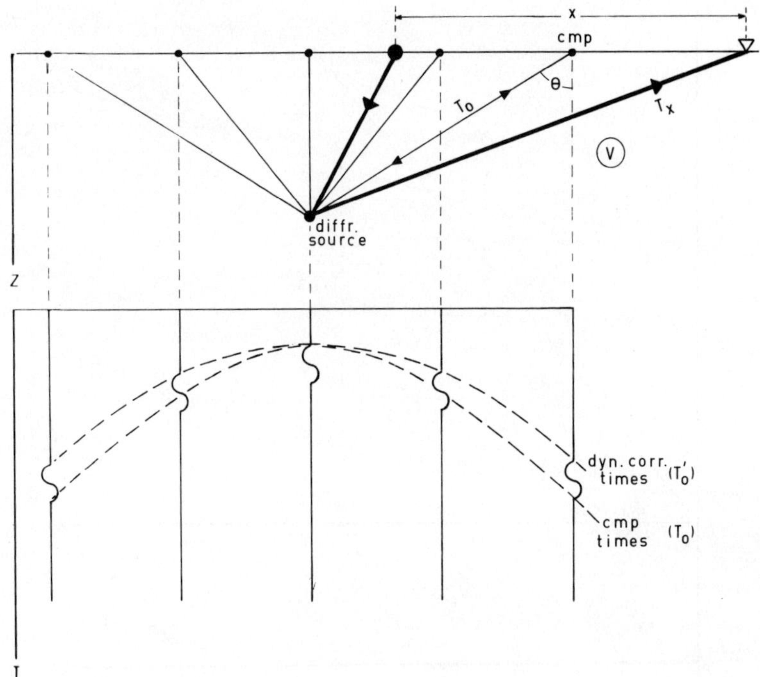

FIG. 3.13. Basic diffraction geometry for c.m.p. data acquisition.

3.3 BASIC REFRACTION GEOMETRY

This section is concerned with a brief review of refraction principles as background to later discussions on continuous velocity logging in wells and the computation of weathering corrections.

Horizontal refractor

Fig. 3.14 shows refraction ray geometry for the case of a single horizontal velocity discontinuity. From wavefront properties the travel time t_{abcd} along the refracted path is equal to the horizontal time between points P and Q in the refracting medium. For a shooting distance x, $t_{abcd} = x/V_2$, so that the total refraction time $t(x)$ is given by

$$t(x) = \frac{x}{V_2} + \frac{2z \cos \theta}{V_1}$$

where θ is the critical angle and z the refractor depth.

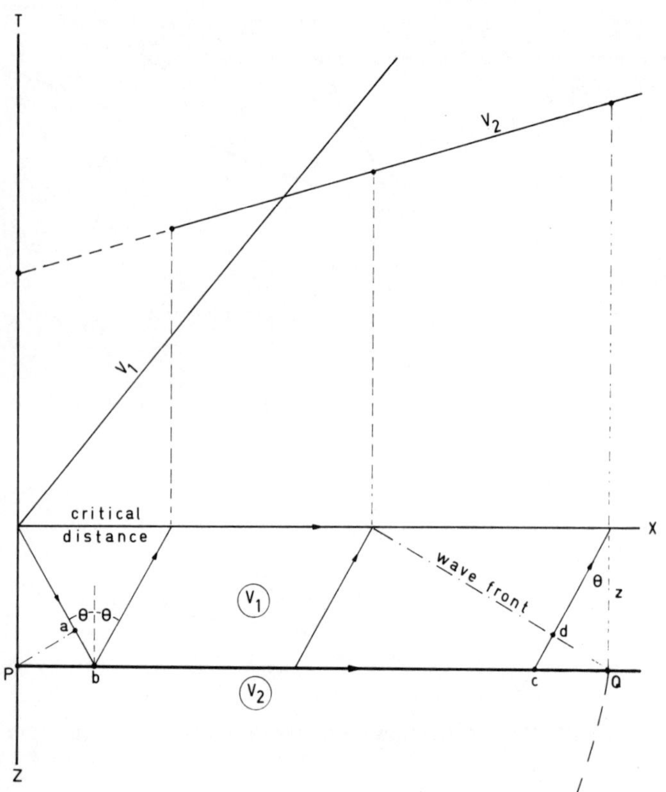

FIG. 3.14. Basic refraction geometry for horizontal interface.

The refraction equation defines a straight line of slope $1/V_2$, which can also be inferred from $dt/dx = (\sin\theta)/V_1$. The V_2 refraction segment commences at the critical distance $2z \tan\theta$ from the shot point. Its backward extrapolation defines the intercept time $(2z \cos\theta)/V_1$, which is also called the delay time for refraction from the V_2 layer.

The t–x segment for the direct arrivals is the line $t(x) = x/V_1$, passing through the shot point. These events can be considered as refractions from the V_1 layer with zero delay time and travel faster than the low velocity surface wave. The intersection of the V_1 and V_2 refraction segments defines the crossover distance, which is always larger than the critical distance.

For the elementary case in Fig. 3.14 the interpretation is straightforward. After measuring velocities along the two refraction segments the refractor depth z can be computed from the graphically derived value for the intercept time.

Inclined refractor

The position of an inclined refractor can be determined from a set of reversed refraction profiles. The relevant refraction ray systems are shown in Fig. 3.15. When the dip angle is δ, emergence angles of refracted raypaths are $(\theta+\delta)$ and $(\theta-\delta)$ for downdip and updip

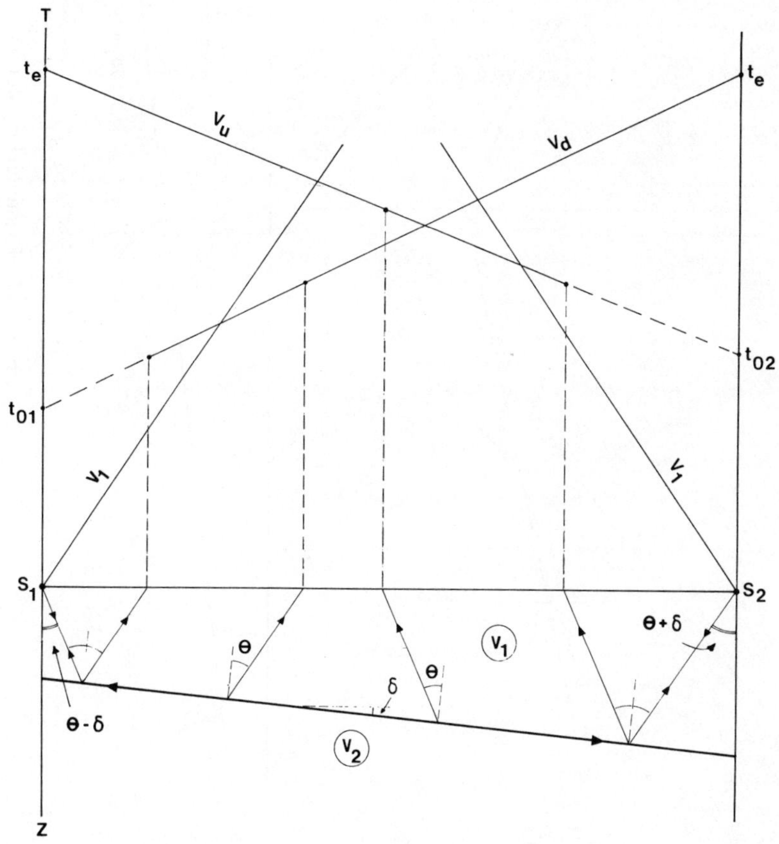

FIG. 3.15. Basic refraction geometry for inclined interface.

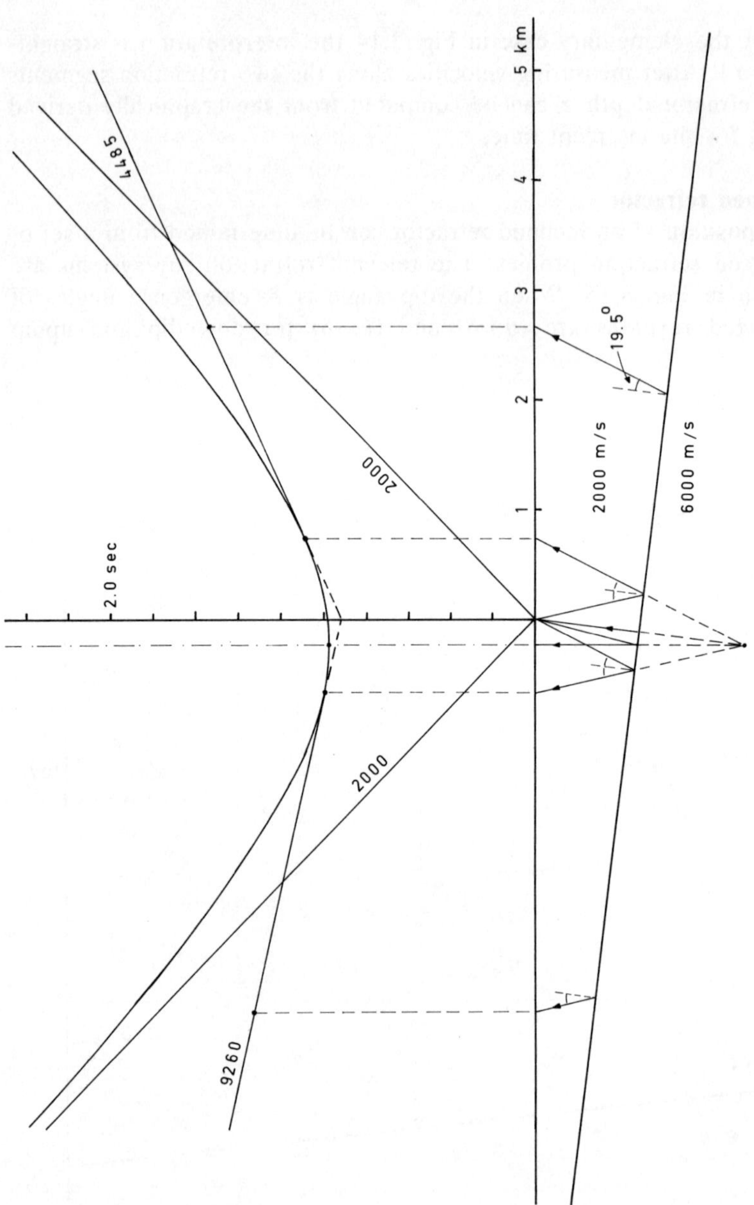

FIG. 3.16. Example 3.5 solution.

profiles respectively. Consequently, the apparent velocity in the downdip direction is given by $V_d = V_1/\sin(\theta+\delta)$ and the one in the updip direction by $V_u = V_1/\sin(\theta-\delta)$. From these relations the critical angle, the dip angle and the refractor velocity can be derived after measuring V_1, V_u and V_d. The refractor velocity is smaller than the average of the two apparent velocities. This can also be inferred from the special case that the dip angle equals the critical angle. In the updip direction the emergence angles are then zero and the apparent velocity infinite. For both profiles the time intercept is $(2z'\cos\theta)/V_1$, z' being the refractor depth below a shot point measured in a direction at right angles to the refracting interface.

Example 3.5 *Reflection–refraction geometry*
Plot reflection and refraction t–x graphs for an interface of 7° dip, having a true depth of 1000 m below the shot point at the centre of a 9000 m spread. Overburden velocity is 2000 m/s, refractor velocity 6000 m/s. Suggested scales: distance and depth 1 cm = 400 m, time 1 cm = 0·2 s.

Example 3.5 *Solution*
From Fig. 3.16 the two high speed refraction segments have a common intercept and are tangent to the reflection hyperbola since at the critical distances reflection and refraction times, and emergence angles of reflected and refracted rays, are equal. Critical distances decrease with decrease of the depth of the velocity interface, reducing the interval of reflection recordings undisturbed by the tails of refraction signals.

Example 3.6 *Critical distance*
Determine the critical distance for a sea bottom reflection when the water depth is 30 m and the sea bed velocity 1650 m/s.

Example 3.6 *Solution*
Critical angle: $\sin^{-1}(1500/1650) = 65{\cdot}4°$.
Critical distance: 60 tan 65·4° = 131 m.

3.4 RAY GEOMETRY IN HORIZONTAL VELOCITY DISTRIBUTIONS

General remarks

Horizontal velocity distributions prevail in clastic basins of Tertiary–Quaternary origin and are occasionally found in restricted parts of older sedimentary systems. Relevant velocity distribution models may consist of sequences of horizontal velocity layers with uniform or nearly uniform velocities, or be represented by continuous velocity–depth functions. In such types of media, reflection t–x curves can always be closely approximated by hyperbolas defined by stacking velocities. Refraction t–x graphs are either smooth entities or composed of a series of linear elements, depending on whether the velocity distribution is of a continuous or discrete nature. Isovelocity planes in continuous distributions are not necessarily conformable to structural bedding planes (see Section 7.5).

In ray tracing problems the direction angle i_n of a ray element in a layer n is defined as the angle it makes with the vertical. The ray parameter p is the ratio of the sine of the ray's direction angle just below the ground surface to the near-surface velocity. From Section 2.10, it is for an ascending ray also equal to the value of the time gradient at its point of emergence. For the special case of a horizontal velocity distribution we have in any velocity layer n from Snell's law that $p = (\sin i_n)/V_n$, V_n being the velocity of layer n. It further obtains from Section 2.3 that all segments of a ray in a horizontal velocity distribution are confined to a common vertical plane.

Discrete distributions

Let p be the parameter of a ray through a sequence of velocity layers with velocities $V_1, V_2, V_3, \ldots$. Where in a layer r the product of p and V_r is equal to one, the ray's direction angle has reached the limiting value of 90°; this corresponds to refraction from layer r. Consequently, when the velocity increases with depth, the theoretical refraction t–x graph will consist of a series of straight line segments having time gradients $1/V_1, 1/V_2, 1/V_3 \ldots$. Fig. 3.17 is an illustrative example, showing a refraction t–x graph for the case of three velocity discontinuities, together with the related set of reflection t–x curves. In general, a refraction segment of slope $1/V_n$ is the asymptote to the reflection t–x curve for reflection from the bottom of layer n, the V_{n+1} refraction segment being tangent to the reflection curve.

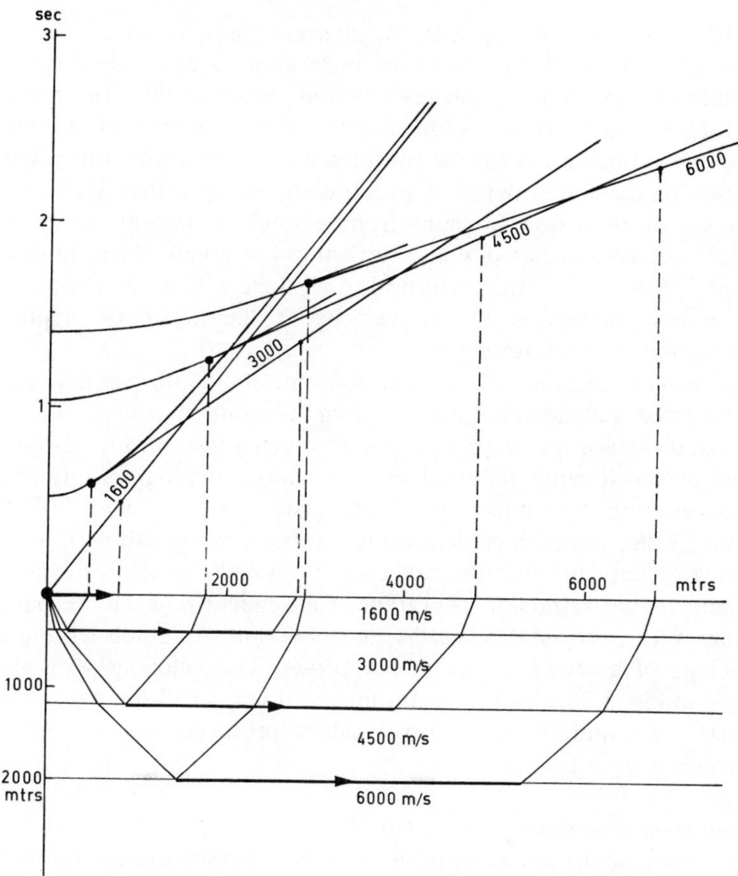

FIG. 3.17. Reflection–refraction geometry for four-layer case.

Continuous distributions

When a velocity distribution is continuous it can often be described by a mathematical expression which leads to practical equations for analysing ray and wavefront geometries. In the early geophysical literature this subject has been covered extensively (Slotnick, 1936a and b) and was referred to as 'the theory of curved paths'. In the present paragraph we shall give a brief outline of some of its principal aspects, with special attention to the linear velocity distribution.

Defining the direction angle i_z of a curved path as the angle between its tangent and the vertical at a depth z, the ray parameter is given by

$p = (\sin i_z)/V_z$, in analogy with the discrete case. When the velocity increases with depth the direction angle along a descending ray will steadily increase until it reaches a limiting value of 90°. The ray will then curve back to the ground surface and emerge at an angle $\sin^{-1}(pV_0)$. Such a ray can be considered as a refracted path; when Z denotes its maximum depth of penetration, we have that $V_Z = 1/p$.

A set of refraction raypaths from a single source at the ground surface can be associated with a refraction t–x graph. Then, in consequence of the above, the reciprocal of its slope will at any point along the x-axis correspond to the velocity at the maximum depth of penetration of a refracted ray.

Let us next consider a horizontal reflector at a depth z and a system of reflected raypaths originating from a common source. For the normal incidence ray the ray parameter is equal to zero. Increasing the value of p will cause reflected rays to emerge at progressively larger distances from the source until the reflection angle becomes 90°. The limiting reflection path is identical to a refraction raypath with parameter $1/V_z$, implying that the reflection t–x graph terminates at and is tangent to the refraction t–x graph. Consequently, when we have a sequence of horizontal reflectors the refraction t–x graph will be the envelope of a set of reflection t–x curves. This relationship is illustrated in Fig. 3.18 which indicates that relatively small source–detector offsets are required for mapping shallow prospects, generalizing the conclusion from Example 3.5.

Fundamental relations

For a horizontal velocity distribution simple parametric equations can be derived which express travel time and horizontal displacement along a ray as a function of the depth z.

From Fig. 3.19, the horizontal displacement is given by

$$x = \int_0^z \tan i_z \, dz$$

whereas for the travel time it obtains that

$$t = \int_0^z \frac{dz}{V_z \cos i_z}$$

Expressing the direction angle i_z in terms of the ray parameter p and

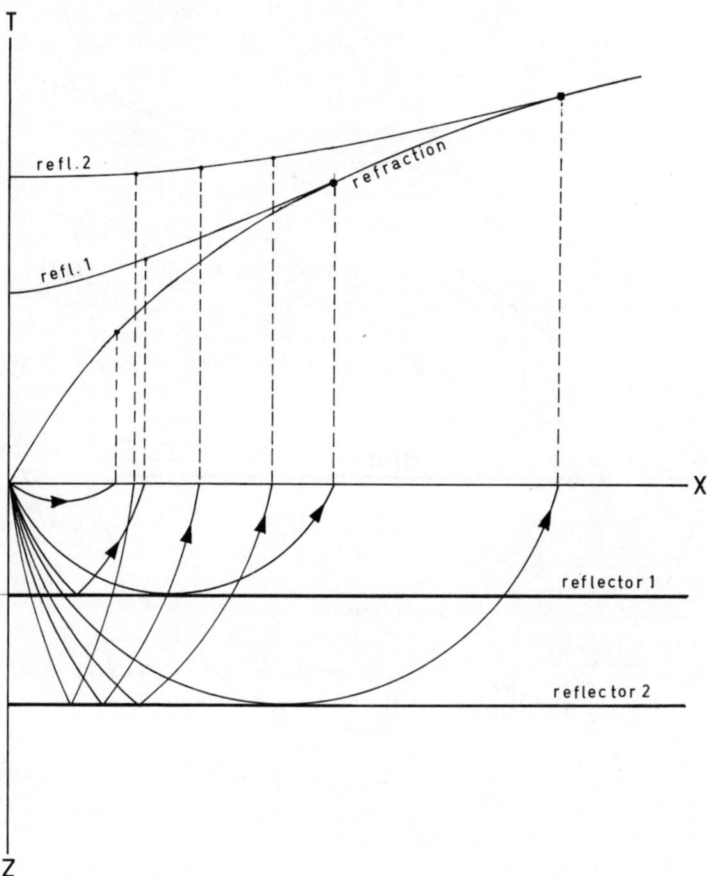

FIG. 3.18. Reflection–refraction geometry in continuous velocity distribution.

the velocity V_z by $p=(\sin i_z)/V_z$, it follows that

$$x(p, z)=\int_0^z \frac{pV_z}{(1-p^2V_z^2)^{\frac{1}{2}}}\,\mathrm{d}z \tag{3.1}$$

and

$$t(p, z)=\int_0^z \frac{\mathrm{d}z}{V_z(1-p^2V_z^2)^{\frac{1}{2}}} \tag{3.2}$$

At the point of emergence of the ray its total horizontal displacement

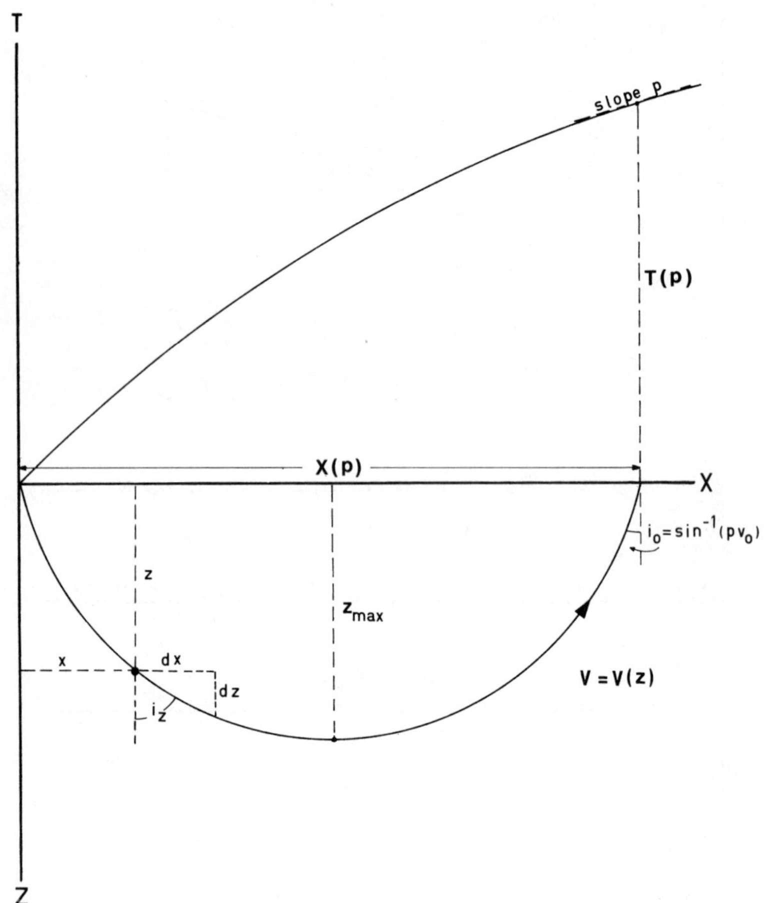

FIG. 3.19. Curved ray geometry; system of symbols.

X and the total travel time T are given by

$$X(p)=2\int_0^Z \frac{pV_z}{(1-p^2V_z^2)^{\frac{1}{2}}}\,dz \tag{3.3}$$

and

$$T(p)=2\int_0^Z \frac{dz}{V_z(1-p^2V_z^2)^{\frac{1}{2}}} \tag{3.4}$$

since the upper integration limit can be expressed as a function of p through $pV_Z=1$.

Upon substitution of a given velocity function the equation of the refraction t–x graph can be derived by elimination of the ray parameter from equations 3.3 and 3.4. Elimination of p from equations 3.1 and 3.2 gives travel time as a function of x and z, which is also the equation of a wavefront curve t in the x–z plane.

The linear velocity distribution

Experience has shown that continuous velocity distributions can usually be closely approximated by a linear function. For the horizontal case it can be written in the form

$$V_z = V_0 + kz$$

which holds in Tertiary sand–shale sections to depths of about 3000 m. V_0 is the initial velocity at datum level and k the velocity gradient. Typical values for V_0 and k in Tertiary clastics are 1600 m/s and $0{\cdot}55\ \mathrm{s}^{-1}$.

An alternative approximation is the Faust relation $V_z = az^b$, for z larger than about 100 m, from which $\log V_z = \log a + b \log z$.

After substitution of the linear velocity function into the integrals 3.1 to 3.4, equations can be derived for raypaths, wavefronts and travel times, showing that rays are circular and wavefronts spherical. These are summarized as follows:

The ray circle:

$$[x - (\cos i_0)/kp]^2 + (z + V_0/k)^2 = 1/k^2p^2 \tag{3.5}$$

Horizontal displacement along ray circle:

$$x = (1/kp)(\cos i_0 - \cos i_z) \tag{3.6}$$

Travel time along ray circle:

$$t = (1/k)[\cosh^{-1}(1/pV_0) - \cosh^{-1}(1/pV_z)] \tag{3.7}$$

Radius of wavefront:

$$r = (V_0/k)\sinh(kt) \tag{3.8}$$

Depth of its centre:

$$z_c = (V_0/k)[\cosh(kt) - 1] \tag{3.9}$$

Vertical time:

$$t(z) = (1/k)\ln(V_z/V_0) \tag{3.10}$$

Vertical depth:

$$z(t) = (V_0/k)(e^{kt} - 1) \tag{3.11}$$

In the above expressions,

$$\sinh u = (e^u - e^{-u})/2$$

$$\cosh u = (e^u + e^{-u})/2$$

and $$\cosh^{-1} u = \ln [u + (u^2 - 1)^{\frac{1}{2}}]$$

Using the Faust relation we find for a vertical time interval

$$t = \frac{1}{a(1-b)} \{z_2^{1-b} - z_1^{1-b}\} \qquad \text{for } z_1 > 100 \text{ m}$$

A horizontal velocity distribution may also be defined by a sequence of linear functions where, for instance, velocity discontinuities are present along unconformities. In that case the depth reference will not change, an initial velocity V_0 associated with a deep velocity unit being a fictive quantity along the earth's surface.

Example 3.7 *Linear velocity distribution*
Given that a raypath is part of a circle with its centre at a height h above the datum plane, show that the velocity function is linear and that $h = V_0/k$.

Example 3.7 *Solution*
From the geometry in Fig. 3.20,

$$\sin i_0/\sin i_z = h/(h + z)$$

From Snell's law,

$$\sin i_0/\sin i_z = V_0/V_z$$

so that

$$V_z = V_0 + zV_0/h = V_0 + kz$$

implying that $h = V_0/k$.

Example 3.8 *Refraction t–x graph*
Given that in a linear velocity distribution rays are circles and wavefronts spheres, show by simple reasoning that the refraction t–x graph

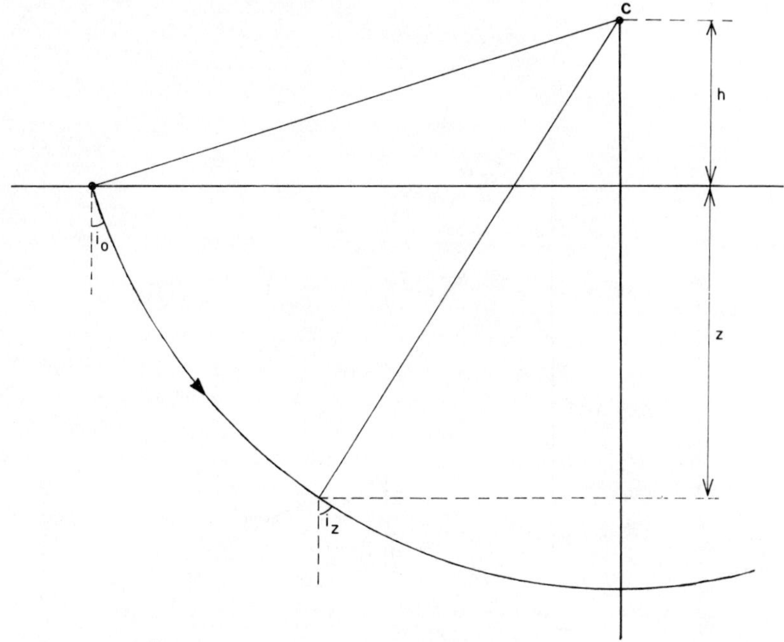

FIG. 3.20. Example 3.7 solution.

can be expressed by the relation

$$t(x) = (2/k)\sinh^{-1}(kx/2V_0)$$

where

$$\sinh^{-1} u = \ln[u + (u^2+1)^{\frac{1}{2}}]$$

Example 3.8 *Solution*
From the geometry in Fig. 3.21,

$$\mathrm{SD} = (V_0^2/k^2 + d^2)^{\frac{1}{2}} - (V_0/k) + d$$

On account of wavefront properties $t_{SZ} = t_{SD}$, so that

$$t_{SZ} = (1/k)\ln[(1 + k^2d^2/V_0^2)^{\frac{1}{2}} + kd/V_0]$$

in consequence of equation 3.10. This is half the refraction time at a distance $2d$ from the source. Upon substitution of $d = \frac{1}{2}x$ and $t_{SZ} = \frac{1}{2}t(x)$, we obtain

$$t(x) = (2/k)\ln[(1 + \tfrac{1}{4}k^2x^2/V_0^2)^{\frac{1}{2}} + \tfrac{1}{2}kx/V_0]$$

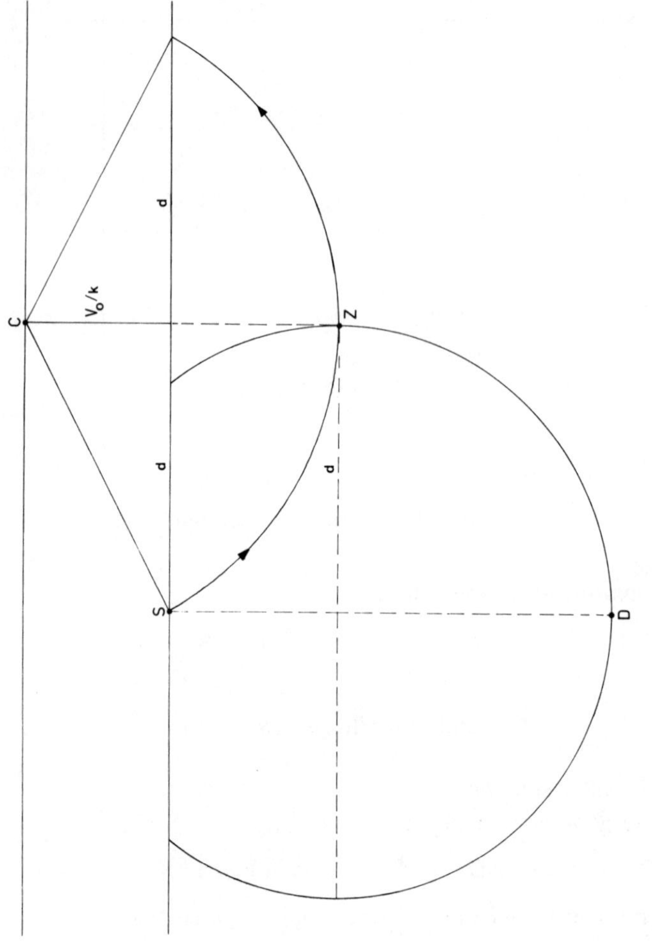

FIG. 3.21. Example 3.8 solution.

which simplifies to

$$t(x) = (2/k) \sinh^{-1}(kx/2V_0)$$

This relation can also be considered as the normal incidence reflection response from a vertical fault plane when plotting the refraction times at half the shooting distances.

Example 3.9 *Two-dimensional ray tracing*
A horizontal velocity distribution is specified as follows

Depth (m)	*Velocity* (m/s)
0–1 200	$V_z = 1\,500 + 0{\cdot}75z$
1 200–1 600	$V_z = 1\,500$
1 600–2 400	$V_z = 1\,500 + 0{\cdot}75z$
<2 400	$V_z = 1\,600 + 0{\cdot}5z$

1. Construct the normal incidence reflection path to a reflection point at 3600 m below the earth's surface and situated on a reflector of 45° dip.
2. Compute the corresponding normal incidence reflection time.

Example 3.9 *Solution*
Lines containing centres of ray circles are at 2000 and 3200 m above the surface. The velocity at 3600 m is 3400 m/s, and the ray parameter is therefore $2{\cdot}08 \times 10^{-4}$ s/m. From $\sin i_z = pV_z$, the direction angles of the ray at the base of the first, second and third velocity layer are 29·9°, 18·2° and 43·3° respectively. From these data the ray in Fig. 3.22 can be constructed. The computation of the reflection time is summarized in the following table:

Layer	V *Top layer*	V *Bottom layer*	*k*	*Two-way time* (s)
1	1 500 m/s	2 400 m/s	0·75	1·370 from equation 3.7
2	1 500	1 500	0	0·561
3	2 700	3 300	0·75	0·685 from equation 3.7
4	2 800	(3 400)	0·50	1·017 from equation 3.7
			Total	3·634

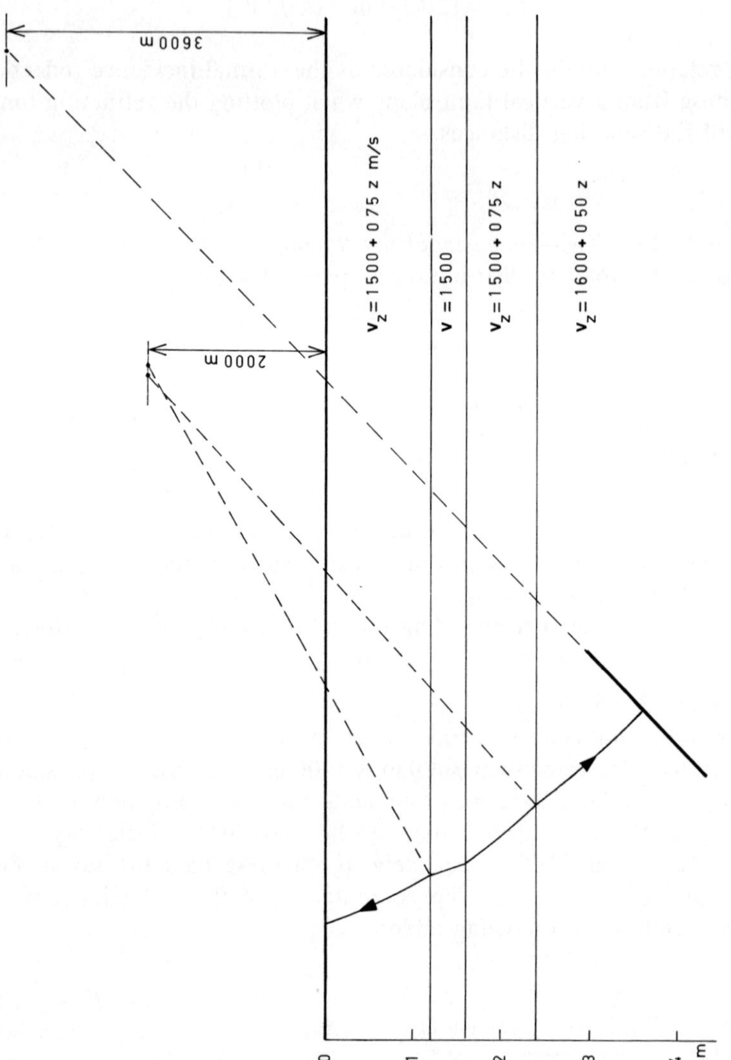

FIG. 3.22. Example 3.9 solution.

3.5 RAY GEOMETRY IN NON-HORIZONTAL VELOCITY DISTRIBUTIONS

Discrete distributions

In non-horizontal velocity distributions dip and strikes of velocity interfaces will generally vary with depth as well as in the lateral direction. Ray segments will then no longer be confined to a common vertical plane. The three-dimensional ray tracing problem can be solved with the aid of vector algebra, representing ray elements by unit vectors expressed in terms of their direction cosines in a rectangular reference frame.

Denoting the direction cosines of unit vectors along incident and transmitted rays by i_1, i_2, i_3 and t_1, t_2, t_3 respectively, and those of the unit vector normal to a velocity interface by n_1, n_2, n_3, and taking from Section 2.3 the relation

$$\frac{\mathbf{I}\times\mathbf{N}}{V_\mathrm{i}}=\frac{\mathbf{T}\times\mathbf{N}}{V_\mathrm{t}}$$

then

$$1/V_\mathrm{i}(n_3i_2-n_2i_3)=1/V_\mathrm{t}(n_3t_2-n_2t_3)$$
$$1/V_\mathrm{i}(n_1i_3-n_3i_1)=1/V_\mathrm{t}(n_1t_3-n_3t_1)$$
$$1/V_\mathrm{i}(n_2i_1-n_1i_2)=1/V_\mathrm{t}(n_2t_1-n_1t_2)$$

Considering that $t_1^2+t_2^2+t_3^2=1$, the solution of these equations gives

$$t_j=\frac{V_\mathrm{t}}{V_\mathrm{i}}\,i_j+\left(\frac{V_\mathrm{t}}{V_\mathrm{i}}\cos\theta_\mathrm{i}-\cos\theta_\mathrm{t}\right)n_j \qquad \text{for } j=1,2,3$$

where

$$\cos\theta_\mathrm{i}=-(i_1n_1+i_2n_2+i_3n_3)$$
$$(\sin\theta_\mathrm{i})/V_\mathrm{i}=(\sin\theta_\mathrm{t})/V_\mathrm{t}$$

Or, in vector notation

$$\mathbf{T}=\frac{V_\mathrm{t}}{V_\mathrm{i}}\mathbf{I}+\left(\frac{V_\mathrm{t}}{V_\mathrm{i}}\cos\theta_\mathrm{i}-\cos\theta_\mathrm{t}\right)\mathbf{N} \qquad \text{(see Appendix 1)}$$

The above relation can be programmed to trace the ray in three dimensions for a given discrete velocity distribution after specifying the upper ray vector. When the x-axis is pointing to the east, and the y-axis to the south, the initial direction cosines are

$$i_1 = -\sin S \sin \theta_0$$
$$i_2 = \cos S \sin \theta_0$$
$$i_3 = \cos \theta_0$$

where θ_0 is the ray's angle of emergence and S is the argument of the relevant time gradient vector measured from the north in the clockwise direction.

Example 3.10 *Three-dimensional ray tracing*
Derive the direction cosines of the transmitted and reflected ray segments associated with the incident ray in the model of Fig. 3.23.

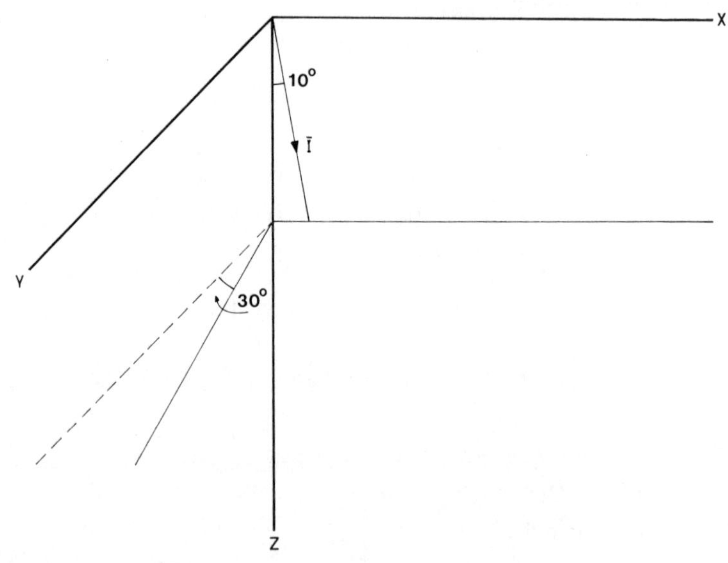

FIG. 3.23. Example 3.10.

Example 3.10 *Solution*
For the incident ray:

$$i_1 = \sin 10° = 0{\cdot}173\,65$$
$$i_2 = 0$$
$$i_3 = \cos 10° = 0{\cdot}984\,81$$

For the normal vector, pointing upward:

$$n_1 = 0$$
$$n_2 = 0{\cdot}5$$
$$n_3 = -0{\cdot}866\ 03$$

Angle of incidence:

$$\cos\theta_i = -\sum i_j n_j = 0{\cdot}852\ 87$$
$$\sin\theta_i = 0{\cdot}522\ 13$$

Angle of transmission:

$$\sin\theta_t = (3/2)0{\cdot}522\ 13 = 0{\cdot}783\ 19$$
$$\cos\theta_t = 0{\cdot}621\ 78$$

Transmitted ray vector:

$$t_1 = 0{\cdot}260\ 48$$
$$t_2 = 0{\cdot}328\ 76$$
$$t_3 = 0{\cdot}907\ 78$$
$$\text{check:}\qquad t_1^2 + t_2^2 + t_3^2 = 1$$

Reflected ray vector: To obtain the direction cosines r_j of the reflected ray replace V_t by V_i, θ_t by $(180-\theta_i)$ and **T** by **R** in the expression for **T**. We then have

$$\mathbf{R} = \mathbf{I} + (2\cos\theta_i)\mathbf{N}$$

from which

$$r_1 = 0{\cdot}173\ 65$$
$$r_2 = 0{\cdot}852\ 87$$
$$r_3 = -0{\cdot}492\ 40$$

Continuous distributions

For a constant velocity gradient a spatial velocity distribution can be represented by the relation

$$V_{xyz} = V_{000} + k_x x + k_y y + k_z z$$

where k_x, k_y and k_z are the velocity gradient components in the directions of the x-, y- and z-axes respectively. In such a velocity model isovelocity planes are parallel and wavefront and ray geometries will be similar to the ones in a horizontal velocity distribution with velocity gradient $(k_x^2+k_y^2+k_z^2)^{\frac{1}{2}}$.

A more general system of inclined but non-parallel isovelocity planes is

$$z = A(V)x + B(V)y + C(V)$$

where the coefficients A, B and C are functions of the velocity V. When results from well velocity surveys indicate that k_z does not change with depth and is thus only dependent on the position of the point of measurement, A, B and C should vary linearly with V. Then we have

$$z = (aV+b)x + (cV+d)y + (eV+f)$$

from which

$$V_{xyz} = (z - bx - dy - f)/(ax + cy + e)$$

which includes the special case of a linear velocity distribution for $a = c = 0$.

A general ray tracing algorithm for continuous velocity distributions can be derived from the one discussed for the discrete case by considering a continuous change of the direction of unit ray vectors, an incident ray vector **I** becoming a transmitted ray vector **I**+d**I** after travelling an infinitesimal distance ds along a curved path. This leads to the following set of equations (Shah, 1973a and b):

$$\frac{\mathrm{d}x}{\mathrm{d}t} = Vi_1 \qquad \frac{\mathrm{d}i_1}{\mathrm{d}t} = \frac{\mathrm{d}V}{\mathrm{d}s}i_1 - \frac{\partial V}{\partial x}$$

$$\frac{\mathrm{d}y}{\mathrm{d}t} = Vi_2 \qquad \frac{\mathrm{d}i_2}{\mathrm{d}t} = \frac{\mathrm{d}V}{\mathrm{d}s}i_2 - \frac{\partial V}{\partial y}$$

$$\frac{\mathrm{d}z}{\mathrm{d}t} = Vi_3 \qquad \frac{\mathrm{d}i_3}{\mathrm{d}t} = \frac{\mathrm{d}V}{\mathrm{d}s}i_3 - \frac{\partial V}{\partial z}$$

where $\mathrm{d}V/\mathrm{d}s = \cos\alpha\,|\mathrm{grad}\ V|$, α being the angle between the vectors **I** and grad V. From these relations a ray can be traced in three dimensions by an integration procedure after specifying its initial direction and the velocity distribution $V(x, y, z)$.

It will be instructive to apply the above equations to the special case of a horizontal velocity distribution. We then have $\partial V/\partial x = \partial V/\partial y = 0$

and $\mathrm{d}V/\mathrm{d}s = i_3(\mathrm{d}V/\mathrm{d}z)$, from which

$$\frac{\mathrm{d}i_1}{\mathrm{d}t} = \frac{\mathrm{d}V}{\mathrm{d}z} i_1 i_3 \qquad \frac{\mathrm{d}i_2}{\mathrm{d}t} = \frac{\mathrm{d}V}{\mathrm{d}z} i_2 i_3 \qquad \frac{\mathrm{d}i_3}{\mathrm{d}t} = \frac{\mathrm{d}V}{\mathrm{d}z} i_3^2 - \frac{\mathrm{d}V}{\mathrm{d}z}$$

Combining these relations with $\mathrm{d}z/\mathrm{d}t = Vi_3$ gives successively

$$\frac{\mathrm{d}i_1}{\mathrm{d}V} = \frac{i_1}{V} \quad \text{and} \quad \frac{\mathrm{d}i_2}{\mathrm{d}V} = \frac{i_2}{V}$$

implying that the ratio i_1/i_2 is constant, so that the ray is confined to a vertical plane, and

$$\frac{\mathrm{d}\sin i_z}{\mathrm{d}V} = \frac{\sin i_z}{V}$$

from which $\sin i_z = pV$, in accordance with Section 3.4.

3.6 REFLECTION TIME DERIVATIVES

From previous sections it appears that there are various ways of expressing reflection time as a function of distance variables. When we consider for instance a common midpoint (c.m.p.) arrangement with u as the c.m.p. coordinate and x the source–receiver offset, $t = f_1(u, x)$ may, for a specific reflector, represent either a c.m.p. record or a 'common x' profile, depending on whether x or u is the independent variable. For the c.m.p. gather the reflection time derivative will be $\partial t/\partial x$, and for the common x gather $\partial t/\partial u$.

Similarly, defining source and receiver positions by distance coordinates s and r we can write for the reflection time $t = f_2(s, r)$. We further have $u = \frac{1}{2}(s + r)$ and $x = r - s$.

From Section 2.10 the reflection time dip $\partial t/\partial r$ along a field record can be expressed in terms of the near-surface velocity and the angle of emergence of a reflected ray. This type of relationship can be derived for all of the above partial derivatives when we consider emergence angles at sources as well as receivers. Denoting these by θ_s and θ_r respectively, we shall analyse two-dimensional near-surface seismic geometry for the following cases:

1. *C.M.P. gathers:* Referring to Fig. 3.24, a small shift $\frac{1}{2}\,\mathrm{d}x$ of the receiver will produce a time increment $(\frac{1}{2}\,\mathrm{d}x \sin\theta_r)/V_0$. Moving the source $\frac{1}{2}\,\mathrm{d}x$ in the opposite direction will contribute

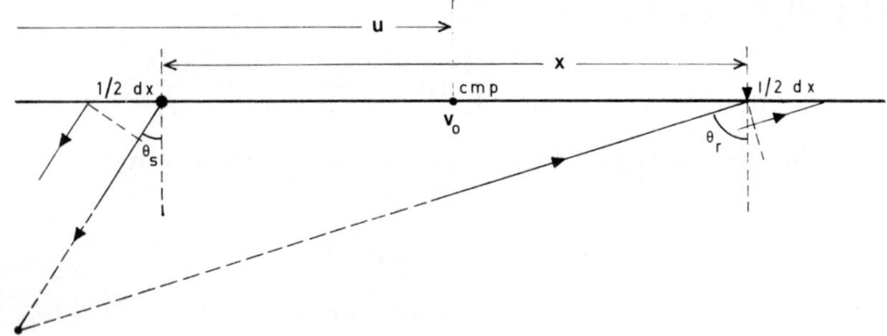

FIG. 3.24. Near-surface reflection geometry for c.m.p. gather.

$-(\frac{1}{2}\,dx \sin \theta_s)/V_0$ to the total time differential. Consequently

$$\frac{\partial t}{\partial x} = \frac{\sin \theta_r}{2V_0} - \frac{\sin \theta_s}{2V_0} \tag{3.12}$$

For a normal incidence ray at the common midpoint $\theta_s = \theta_r$. Therefore, when $x = 0$ we have $\partial t/\partial x = 0$, from which we can be assured that the t–x graph for primary reflection events on a c.m.p. gather will terminate at right angles to the time axis, irrespective of the complexity of the velocity distribution. It should be stressed that this is only true when the normal incidence condition is fulfilled. Otherwise a non-zero time dip at the zero offset will be obtained from the relation just derived.

2. *Common offset gathers:* From Fig. 3.25, for a small shift du of

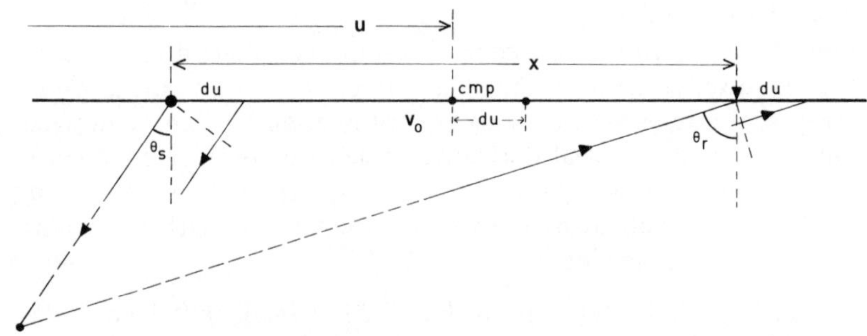

FIG. 3.25. Near-surface reflection geometry for common offset gather.

source and receiver in the same direction we have

$$dt = (du \sin \theta_s) V_0 + (du \sin \theta_r) V_0$$

so that

$$\frac{\partial t}{\partial u} = \frac{\sin \theta_s}{V_0} + \frac{\sin \theta_r}{V_0} \tag{3.13}$$

For the special case that $x = 0$, it obtains that $\theta_s = \theta_r = i_0$. We can then write $\partial t/\partial u = (2 \sin i_0)/V_0$, which is the time dip along a normal incidence reflection section.

3. *Reflection spreads:* With reference to Fig. 3.26 we have for a

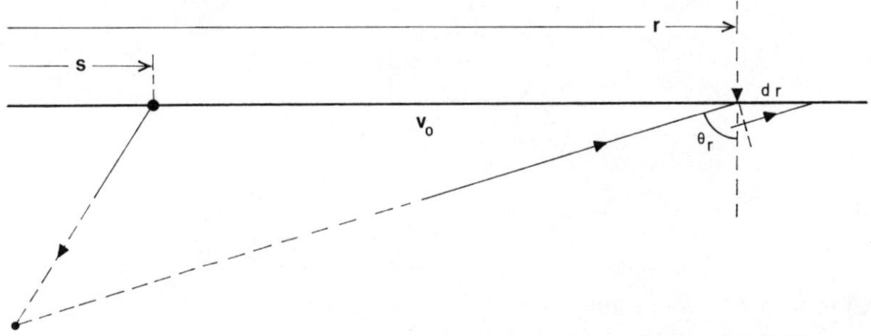

FIG. 3.26. Near-surface reflection geometry for single spread.

reflection spread $t = f_2(r, s)$. Keeping s fixed, it is recalled that

$$\frac{\partial t}{\partial r} = \frac{\sin \theta_r}{V_0} \tag{3.14}$$

By combination of expressions 3.12, 3.13 and 3.14 we further have

$$\frac{\partial t}{\partial r} = \frac{\partial t}{\partial x} + \frac{1}{2}\frac{\partial t}{\partial u}$$

and

$$\left(\frac{\partial t}{\partial u}\right)_{x=0} = 2\left(\frac{\partial t}{\partial r}\right)_{r=s}$$

The above two-dimensional approach can be extended to the general case of arbitrarily oriented traverses by relating time dips to emergence angles as well as to the direction of a traverse with reference to that of time gradients (Section 2.10). This will not alter the previous conclusions.

Example 3.11 *Reflection time derivatives*

For a reflection event along a stacked profile normal incidence times are given by

$$t = 10^{-6}u^2 + 1{\cdot}990 \quad \text{s} \tag{3.15}$$

On the c.m.p. gather for which $u = 100$ m, the corresponding t–x relation is

$$t^2(u, x)_{u=100} = t^2(u, 0)_{u=100} + x^2/3000^2 \quad \text{s}^2 \tag{3.16}$$

Determine:

1. $$t(u, 0)_{u=100}$$
2. $$\left(\frac{\partial t}{\partial u}\right)_{u=100}$$
3. $$\left(\frac{\partial t}{\partial x}\right)_{x=0} \quad \text{and} \quad \left(\frac{\partial^2 t}{\partial x^2}\right)_{x=0}$$
4. $$\left(\frac{\partial t}{\partial r}\right)_{s=r}$$

Example 3.11 *Solution*

1. From equations 3.15 and 3.16

$$t(u, 0)_{u=100} = t_{u=100} = 2{\cdot}0 \quad \text{s}$$

2. From equation 3.15,

$$\left(\frac{\partial t}{\partial u}\right)_{u=100} = 2 \times 10^{-4} \quad \text{s/m}$$

3. From equation 3.16,

$$\left(\frac{\partial t}{\partial x}\right)_{x=0} = 0 \quad \text{and} \quad \left(\frac{\partial^2 t}{\partial x^2}\right)_{x=0} = 5{\cdot}6 \times 10^{-8} \quad \text{s/m}^2$$

4. From

$$\frac{\partial t}{\partial r} = \frac{\partial t}{\partial x} + \frac{1}{2}\frac{\partial t}{\partial u}$$

we get

$$\left(\frac{\partial t}{\partial r}\right)_{s=r} = \left(\frac{\partial t}{\partial x}\right)_{x=0} + \frac{1}{2}\left(\frac{\partial t}{\partial u}\right)_{u=100} = 10^{-4} \quad \text{s/m}$$

3.7 SPURIOUS *t–x* RELATIONS

Spurious events are associated with ray systems which contain divergent up- and downgoing elements at the common midpoint. From Section 3.6 their *t–x* relations on c.m.p. gathers will have a non-zero derivative at the zero offset. Spurious signals have therefore generally

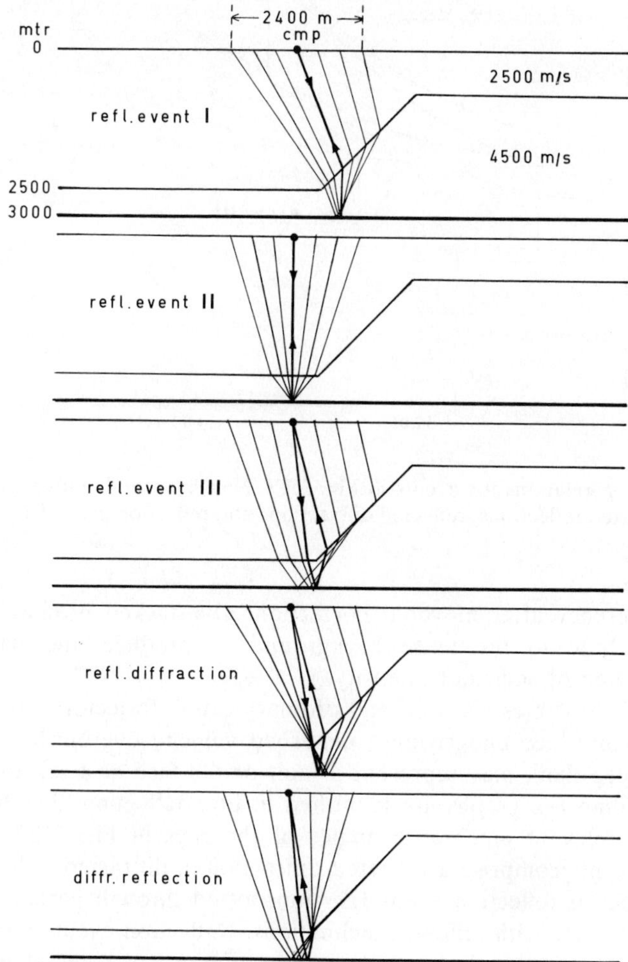

FIG. 3.27. Common midpoint ray geometries for various types of seismic events.

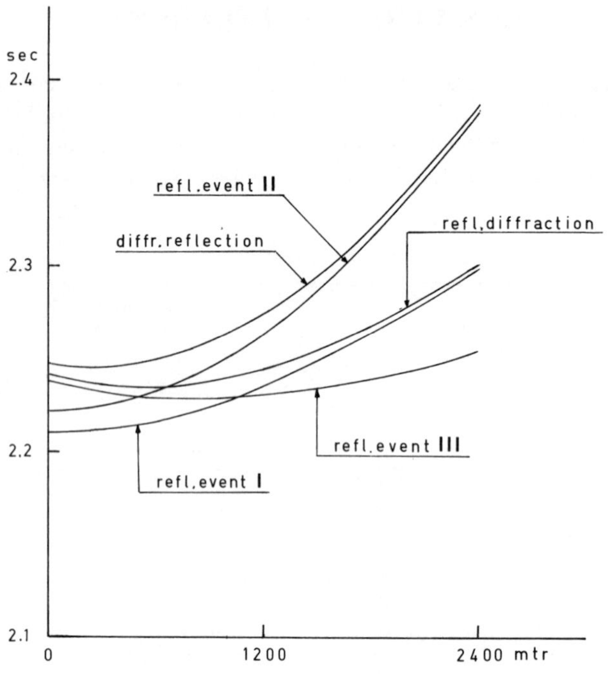

FIG. 3.28. *t–x* relations for events in Fig. 3.27. Non-hyperbolic moveouts for diffracted reflections, reflected diffractions and reflection event III.

a low coherency after moveout correction. On stacked profiles they may contribute to the noise background or produce unexplained artefacts when of sufficient intensity.

Fig. 3.27 illustrates aberrant and ordinary c.m.p. trajectories from a horizontal interface underlying a disturbed velocity overburden. The model's steep flank may represent a fault or the face of a salt diapir. Corresponding *t–x* graphs are identified in Fig. 3.28, time dips being related to relevant emergence angles of the rays in Fig. 3.27. The spurious events comprise a reflected diffraction, a diffracted reflection and the special reflection event III, transmitted through parts of the velocity interface with different inclinations. Reflection events I and II are of the normal type but would require different stacking velocities. This will cause one of these reflections to be attenuated on a stacked profile.

3.8 VELOCITY DISTRIBUTION SUBSTITUTES

Velocity distribution substitutes are empirically derived stacking velocities and mathematically defined quantities associated with a horizontal velocity distribution. Under favourable conditions estimates of interval velocities may be obtained from stacking velocities. This usually provides the only source of velocity information during reflection interpretation in regions devoid of well control.

In a horizontal velocity distribution we may distinguish between average velocities, mean velocities and root-mean-square (r.m.s.) velocities.

The average velocity is defined as the ratio of vertical depth to corresponding vertical time, or

$$V_{av} = \frac{\int_0^z dz}{\int_0^z \frac{dz}{V_z}}$$

Using an average velocity as a substitute for a horizontal velocity distribution is equivalent to replacing a least time reflection path by a pair of straight lines. Therefore, a hyperbolic approximation of the reflection t–x graph from a horizontal reflector of the form

$$t_x^2 = t_0^2 + x^2/V_{av}^2$$

will be delayed with respect to the observed t–x relation, except at the zero offset.

The mean velocity is the mathematical mean of the velocity function. It is given by

$$V_m = \frac{\int_0^z V_z \, dz}{\int_0^z dz}$$

The r.m.s. velocity is the square root of the mean of the squared velocities taken as a function of two-way vertical time, from which

$$V_{rms}^2 = \frac{\int_0^{t_0} V^2(t_0) \, dt_0}{t_0}$$

In view of $dz = \frac{1}{2}V(t_0)\,dt_0$ and $t_0 = \frac{1}{2}\int_0^z dz/V_z$, the above expression can also be written in the form

$$V_{rms}^2 = \frac{\int_0^z V_z\,dz}{\int_0^z \frac{dz}{V_z}}$$

so that $V_{rms}^2 = V_m V_{av}$.

It should be noted that V_{rms} is not a ray tracing or depth conversion velocity. Therefore, two-way vertical time is not identical to $2z/V_{rms}$; correctly, however, $t_0 = 2z/V_{av}$ from the definition of the average velocity.

The seismic significance of V_{rms} follows from the approximation of the observed reflection t–x relation by a Maclaurin series expansion. From Section 3.6 we can write for a normal incidence reflection event on a c.m.p. gather

$$t_x^2 = f(x^2)$$

Taking x^2 as the independent variable, the squared reflection time in the form of a Maclaurin power series becomes

$$t_x^2 = t_0^2 + \frac{x^2}{1!}f^{(1)}(x^2)_0 + \frac{x^4}{2!}f^{(2)}(x^4)_0 + \ldots$$

where $f^{(n)}(x^2)$ represents the nth order derivative of $f(x^2)$ with respect to x^2.

Neglecting the third and higher order terms, the hyperbolic approximation to the reflection t–x graph becomes

$$t_x^2 = t_0^2 + x^2 f^{(1)}(x^2)_0$$

The coefficient of x^2 in this relation can be evaluated by considering that

$$f^{(1)}(x^2) = \frac{dt_x^2}{dx^2} = \frac{dt_x}{dx}\frac{t_x}{x}$$

where for a *horizontal reflector* $dt_x/dx = (\sin i_0)/V_0 = p$, the ray parameter. Then, from equations 3.1 and 3.2 it obtains for x and p approaching zero that

$$f^{(1)}(x^2)_0 = \frac{\int_0^z V_z\,dz}{\int_0^z \frac{dz}{V_z}} = \frac{1}{V_{rms}^2}$$

which shows that, for zero reflector dip, V_{rms}^2 is the inverse slope of a $t_x^2 - x^2$ plot at its origin.

For an arbitrary velocity distribution the *stacking velocity* V_s gives the best fit of hyperbolic moveouts derived from $t_x^2 = t_0^2 + x^2/V_s^2$ to the actual reflection moveouts.

For the horizontal case the relationship between all of the above velocity quantities follows from Fig. 3.29, showing the $t_x^2 - x^2$ relation for a recorded reflection on a c.m.p. gather and straight $t_x^2 - x^2$ plots for hyperbolic approximations based on V_{av}, V_m, V_{rms} and V_s. From the Fermat principle reflection times based on average velocities are always too large, whereas according to the definition of V_s the line $t_x^2 = t_0^2 + x^2/V_s^2$ should cross the observed $t_x^2 - x^2$ relation. Since V_{rms}^2 is its inverse slope at the origin we have $V_{av} < V_{rms} < V_s$. Shortening the

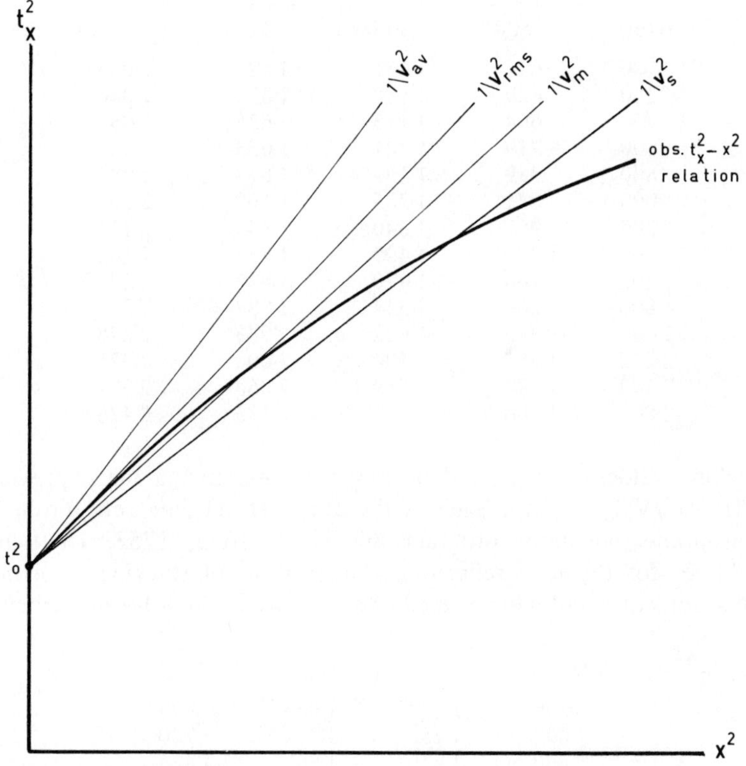

FIG. 3.29. Approximations to true $t_x^2 - x^2$ relation of reflection event by means of various velocity distribution substitutes.

spread will reduce the difference between V_s and V_{rms} so that the limit to which V_s tends when x approaches zero is identical to the r.m.s. velocity. This suggests that in practice we may for recordings from a horizontal reflector sequence and for moderate source–receiver offsets assume that $V_s \approx V_{rms}$. From $V_{rms}^2 = V_{av} V_m$ we have $V_m > V_{rms}$ so that the stacking velocity may occasionally be a close measure of the mean velocity.

Illustrative example

For a velocity distribution $V_z = 1500 + 0{\cdot}5z$ and reflectors at 500, 1000, 1500 and 2000 m, (t_x, x) pairs from equations 3.6 and 3.7 are tabulated below.

	t_x (ms)			
x (m)	$z = 500$	$z = 1\,000$	$z = 1\,500$	$z = 2\,000$
0	617	1 151	1 622	2 043
200	629	1 156	1 625	2 046
400	664	1 173	1 636	2 053
600	719	1 201	1 653	2 066
800	789	1 239	1 678	2 083
1 000	871	1 285	1 708	2 105
1 200	962	1 340	1 745	2 131
1 400	1 059	1 402	1 787	2 162
1 600	1 160	1 470	1 835	2 197
1 800	1 266	1 544	1 887	2 236
2 000	1 373	1 622	1 943	2 278
2 200	1 483	1 704	2 004	2 325
2 400	1 594	1 789	2 068	2 374
2 600	1 706	1 877	2 135	2 426

Stacking velocities can be determined by minimizing the expression $\sum[(t_x^2 - x^2/V_s^2)^{\frac{1}{2}} - t_0]^2$ for each of the above (t_x, x) data sets. From an appropriate calculator program this yields 1632, 1753, 1870 and 1984 m/s for the four reflectors. Comparisons of stacking velocities, r.m.s. velocities and mean velocities are shown in the following table:

Depth	V_s	V_{rms}	V_m
500	1 632	1 623	1 625
1 000	1 753	1 744	1 750
1 500	1 870	1 862	1 875
2 000	1 984	1 979	2 000

N.M.O. velocities

When the reflector is inclined the inverse slope of its $t_x^2 - x^2$ plot at the zero offset is usually referred to as V_{nmo}^2. From previous considerations it is generally true that

$$1/V_{\mathrm{nmo}}^2 = f^{(1)}(x^2) = \frac{\mathrm{d}t_x}{\mathrm{d}x}\frac{t_x}{x} \qquad \text{for } x = 0$$

For the special case of a horizontal velocity distribution V_{nmo} may be related to the reflector dip δ. With reference to Fig. 3.30 and Section

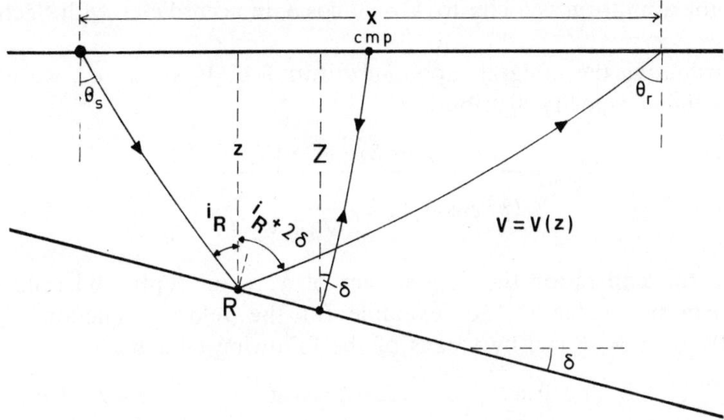

FIG. 3.30. Derivation of V_{nmo} for inclined reflector in horizontal velocity distribution.

3.6 this involves an evaluation of the derivative $\mathrm{d}t_x/\mathrm{d}x$ from

$$\frac{\mathrm{d}t_x}{\mathrm{d}x} = \frac{\sin\theta_{\mathrm{r}}}{2V_0} - \frac{\sin\theta_{\mathrm{s}}}{2V_0}$$

with $\sin\theta_{\mathrm{s}} = -(V_0/V_{\mathrm{R}})\sin i_{\mathrm{R}}$ and $\sin\theta_{\mathrm{r}} = (V_0/V_{\mathrm{R}})\sin(i_{\mathrm{R}} + 2\delta)$, V_{R} being the velocity at the level of the reflection point R. Then, after determining the ratio t_x/x from equations 3.1 and 3.2, $f^{(1)}(x^2)_0$ can be derived through a limiting procedure with the aid of L'Hôpital's rule (see Appendix 2). This leads to

$$V_{\mathrm{nmo}}^2 = \frac{\displaystyle\int_0^Z \frac{V_z\,\mathrm{d}z}{Q_z}}{\displaystyle\int_0^Z \frac{\mathrm{d}z}{V_z Q_z}} + p^2 \frac{\displaystyle\int_0^Z \frac{V_z^3\,\mathrm{d}z}{Q_z^3}}{\displaystyle\int_0^Z \frac{\mathrm{d}z}{V_z Q_z}}$$

where the upper integration limit Z is the depth of the normal incidence reflection point and $Q_z = (1 - V_z^2 p^2)^{\frac{1}{2}}$, p being the parameter of the normal incidence ray. It can easily be verified that for a zero reflector dip the above relation reduces to

$$V_{\rm rms}^2 = \frac{\int_0^Z V_z \, dz}{\int_0^Z \frac{dz}{V_z}}$$

and for a uniform velocity to $V_s = V/\cos\delta$, in accordance with Section 3.1.

Evaluating the integrals upon substitution of $V_z = V_0 + kz$, we have for a linear velocity distribution

$$V_{\rm nmo}^2 = \frac{(\cos\delta) - (1 - V_0^2 p^2)^{\frac{1}{2}}}{(p^2 \cos\delta) \ln V_0 \dfrac{1 + \cos\delta}{V_R + V_R(1 - V_0^2 p^2)^{\frac{1}{2}}}}$$

From this expression the dependence of $V_{\rm nmo}$ on depth and reflector dip can be evaluated. An example for the velocity function $V_z = 1500 + 0{\cdot}5z$ is shown by means of the following tabulation:

	$Z = 1\,000$ m		$Z = 1\,500$ m		$Z = 2\,000$ m	
δ	$V_{\rm nmo}$	$V_{\rm rms}/\cos\delta$	$V_{\rm nmo}$	$V_{\rm rms}/\cos\delta$	$V_{\rm nmo}$	$V_{\rm rms}/\cos\delta$
0	1 744	1 744	1 862	1 862	1 979	1 979
10	1 758	1 771	1 877	1 891	1 995	2 009
20	1 800	1 856	1 923	1 981	2 045	2 106
30	1 877	2 014	2 007	2 150	2 135	2 285
40	1 999	2 277	2 139	2 431	2 278	2 583
50	2 188	2 713	2 345	2 897	2 501	3 079
60	2 490	3 488	2 675	3 724	2 858	3 958
70	3 031	5 099	3 266	5 444	3 497	5 786
80	4 300	10 043	4 649	10 723	4 994	11 397

These data indicate that the approximation $V_{\rm nmo} \approx V_{\rm rms}/\cos\delta$, in analogy with $V_s = V/\cos\delta$, appears only valid for small to moderate dips. For the present case a fairly accurate description of $V_{\rm nmo}$ values can be obtained from the relation $V_{\rm nmo} \approx 2V_{\rm rms}/(1 + \cos\delta)$, for dips up to about 40°.

The above is a two-dimensional analysis. From Section 3.1, $V_{\rm nmo}$ is a direction dependent quantity. In non-horizontal velocity distributions its value will further depend on the velocity regimes of the individual velocity units and the spatial orientations of velocity interfaces.

In severely disturbed velocity overburdens reflection moveouts may deviate considerably from hyperbolic moveouts. This causes a distortion or a dimming of stacked reflection information and a shift of the stacked signal from its theoretical normal incidence position which may vary from trace to trace. These difficulties may be resolved by pre-stack migration, which is not concerned with a dynamic correction procedure (Chapter 8).

Interval velocities

From the definition of V_{rms} the r.m.s. interval velocity is given by

$$V_i^2 = \frac{\int_{t_1}^{t_2} V^2(t_0)\,\mathrm{d}t_0}{t_2 - t_1}$$

the integration limits referring to the two-way vertical time to the top and to the base of the layer under consideration. After expanding the integral it obtains that

$$V_i^2 = \frac{\bar{V}_2^2 t_2 - \bar{V}_1^2 t_1}{t_2 - t_1} \qquad \text{(Dix's formula)}$$

$\bar{V}_1$ and $\bar{V}_2$ designate r.m.s. velocities for the intervals between the ground surface and the top and the base of the layer respectively. For a layer thickness of less than about 500 m and velocity distributions normally encountered the r.m.s. velocity is nearly equal to the average interval velocity. For the special case that $V_s \approx V_{rms}$, average velocities along restricted subsurface intervals can then be estimated from the Dix relation. For the data set of the Illustrative Example in this section, average interval velocities, r.m.s. interval velocities and their estimates from the stacking velocity distribution are listed as follows:

Interval	V_{av}	V_{rms}	V_{rms} *from* V_s
0–500	1 622	1 623	1 632
500–1 000	1 872	1 874	1 883
1 000–1 500	2 122	2 124	2 129
1 500–2 000	2 373	2 374	2 373

For non-horizontal velocity layering various authors have dealt with the problem of estimating interval velocities from the variation of V_{nmo} values with normal incidence reflection time. Shah (1973a and b) developed a two-dimensional procedure, considering layers of uniform velocity separated by interfaces of arbitrary dips. An extension of Shah's algorithm to the three-dimensional case has been reported by Krey (1976) and Hubral (1976). The practical application of these methods is restricted to simple velocity structures and dependable (V_{nmo}, t_0) data sets.

Velocity analysis techniques

Velocity analysis is the determination of stacking velocities as a function of zero-offset recording time from moveout variations on a c.m.p. gather. Velocity check points may be selected from reflection information on provisionally processed profiles, the brute stacks. They are normally separated by intervals of the order of 2 to 5 km, along which stacking velocity distributions for consecutive c.m.p.s can be obtained by interpolation.

A simple velocity analysis scheme is the *velocity scan*, based on visual inspection of the coherency of dynamically corrected traces for a series of trial velocities which are kept constant down the c.m.p. record and incremented by small steps. The corrected gathers are displayed side by side in order of increasing velocities and the stacking velocity distribution determined by the tracing of straight signal alignments. Similarly, stacking velocities can be derived by searching for amplitude maxima along a sequence of *constant velocity stacks* from a single c.m.p. gather.

A more elaborate procedure involves the quantification of the coherency of moveout corrected events in a time window having a duration of about 20 to 40 ms and which is moved along the c.m.p. record in steps of the order of 20 ms. For each window position, defined by that of its centre, the multi-channel coherency is expressed in terms of a cross-correlation or semblance coefficient (Al-Chalabi, 1979). These are statistical quantities whose significance increases with the increase of the multiplicity of the reflection coverage.

Velocity analysis results obtained by this method are usually presented in the form of listings of coherency values in a $t_0 - V_s$ reference frame. These may also be recorded as variable density plots or graphs showing the coherency as a function of velocity for each position of the time window. Displays of this type are referred to as *velocity spectra* and the individual coherency build-ups as events.

The precision of stacking velocity measurements decreases generally with depth by noise interference and a reduced sensitivity of moveout values to variations of the stacking velocity. This follows from

$$d(\Delta t)_x = -(x^2/t_x V_s^3)\, dV_s$$

which can be derived from the moveout equation by differentiation. The above also implies that there is no need for highly accurate stacking velocities at the deeper recording levels.

The interpretation of velocity spectra involves the discrimination of coherency maxima due to primary reflections against the ones generated by multiple reflections, diffractions, fault plane reflections and spurious signal alignments. Relatively high velocities may be associated with the steep dip events and low velocities with multiple reflections. The distribution of coherency maxima should be studied with close reference to the provisional stack. This enables the anticipation of changes of direction of the $t_0 - V_s$ trend where major reflection events suggest corresponding changes of velocity regimes. An additional aid is a plot of coherency values as a function of t_0 and $t_0 V_s^2$. From the Dix relation this may reveal alignments of coherency maxima which correspond to squared formation interval velocities.

Owing to signal stretch (Section 3.1) maximum coherency values will be reached after a small overcorrection of reflection moveouts. There will therefore be a tendency to underestimate the theoretical stacking velocities during the above described procedure.

An alternative velocity analysis scheme is the *moveout scan.* It uses the reflection moveout at the maximum offset, $(\Delta t)_X$, as the primary variable, which is incremented in steps of about 10 ms within the expected seismic range. For each value of $(\Delta t)_X$ the moveout formula $(\Delta t)_X = (t_0^2 + X^2/V_s^2)^{\frac{1}{2}} - t_0$ defines a $t_0 - V_s$ distribution from which a corresponding set of $(\Delta t)_x$ values can be determined. From Section 3.1 $(\Delta t)_x \approx (\Delta t)_X (x^2/X^2)$, so that the moveout correction for any shot–detector offset will be nearly uniform, minimizing signal stretch. Varying $(\Delta t)_X$ will produce power maxima along a sequence of stacked gathers at various time levels. The relationship between t_0 and $(\Delta t)_X$ can then be established and the stacking velocity distribution derived from the above expression for $(\Delta t)_X$. Fig. 3.31 is a graphical illustration of this principle.

More advanced velocity determination systems exercise full control on stacking velocity variations in the lateral direction by performing velocity analysis procedures at all or at a set of closely spaced common midpoints. The massive amount of data that is collected in that manner requires special data handling techniques which scrutinize event

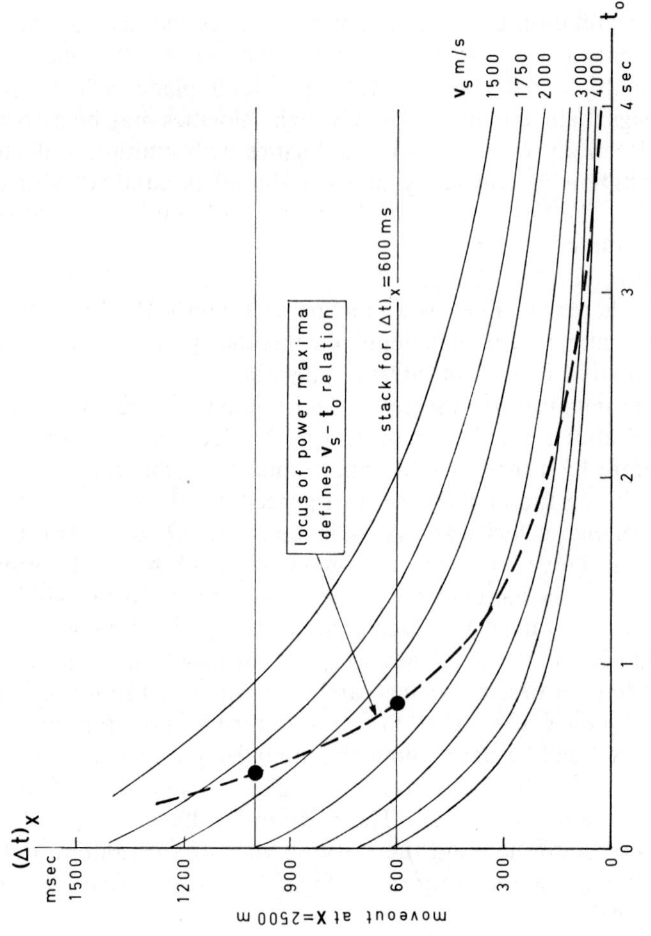

FIG. 3.31. Illustration of moveout scan procedure.

characteristics for lateral consistency. These are based on the criterion that representative power or coherency maxima and associated quantities such as time, velocity and moveout should be repetitive at consecutive c.m.p.s within preset limits and along a prescribed lateral interval. This is referred to as event tracking, the inconsistent events being removed from the data set. An additional feature is the regular updating of thresholds during the tracking procedure.

Reflection time dip may also be used as a search parameter. Time dips related to a $t_0 - V_s$ or a $t_0 - (\Delta t)_X$ event panel can be analysed by subjecting a limited number of contiguous panels on both sides of it to small static shifts which compensate for an assumed time dip expressed in milliseconds per c.m.p. interval. Centre and side panels are subsequently stacked, enhancing coherency or power maxima where the actual reflection time dip matches the assumed one. The time dip is incremented by small steps and the process repeated until the expected dip range has been covered. This adds time dip to the event characteristics already mentioned, so that they are now time, time dip, velocity (or moveout) and coherency (or power) which can be examined for lateral consistency. Eventually a file is created which contains velocity information pertaining to the accepted multiple and primary reflections. From this a primary stacking velocity distribution can be derived for all c.m.p. gathers along the reflection traverse.

3.9 MULTIPLE REFLECTIONS

A multiple reflection is a reflection event that is produced by two or more successive reflections from a combination of layer boundaries, including the earth's surface.

From a general subdivision of the various types of multiple ray geometries a distinction can be made between long-path and short-path multiples, also referred to as long-delay and short-delay multiples. For a given interface the arrival time of its long-path multiple may be twice that of the primary reflection. At a deeper recording level the long-path multiple may then be mistaken for a primary event when it is conformable to primary reflections. Short-delay multiples follow primaries at short time intervals. They are not distinguishable as separate events and may either interfere with the primary reflection or lengthen its tail, impairing vertical resolution.

A simple short-path multiple is the two-bounce reflection; it is the ghost from reflection of the outgoing signal against the earth's surface or the base of the weathered layer. Multiples which are produced by three reflections are called first order multiples, nth order multiples being associated with $(2n+1)$ reflections from a group of layer interfaces. In view of the generally small values of reflection coefficients in the subsurface, the amplitudes of multiple reflections diminish rapidly with an increase of their order. Therefore, of all the long-path multiples it is usually the first and second order surface multiples and the first order internal multiples which can compete with primary reflections. A summary of the geometry and generally accepted nomenclature for various types of multiple configurations is shown in Fig. 3.32.

The attenuation of long-path multiples during c.m.p. stacking is based on the differences between the moveouts of multiples and primaries at identical recording levels. Moveouts associated with primary reflections are referred to as normal moveouts, from which the primary stacking velocity distribution is derived during velocity analysis.

For horizontal velocity layering we have the simplifying condition that at the c.m.p. up- and downgoing multiple rays are coincident so that, from Section 3.6, the travel time curve for the multiple event can be approximated by a symmetrical hyperbola. Depending on whether the formation velocity increases or decreases with depth, the multiple stacking velocity will then either be smaller or larger than the stacking velocity for a primary event of the same travel time. When the velocity in a horizontal reflector sequence is uniform, however, multiple reflections cannot be distinguished from primary events.

A full discussion of multiple reflection geometry must take account of reflector dip. The moveout of a multiple from an inclined interface will be smaller than that in the absence of dip under otherwise similar conditions. It is then conceivable that it may be less than the moveout of a primary reflection from a deeper and lower dipping horizon, even when the velocity is uniform or decreasing with depth. We shall illustrate this condition for a first order W-type multiple whose trajectory is confined between the ground surface and an interface for which the dip will be varied. The relevant model is shown in Fig. 3.33. For a fixed vertical depth of 1000 m of the shallow horizon below the c.m.p. and for dip angles of zero, five and ten degrees the positions of the deeper horizontal reflector have been chosen to yield primary reflections with zero-offset times which are identical to those of the multiple

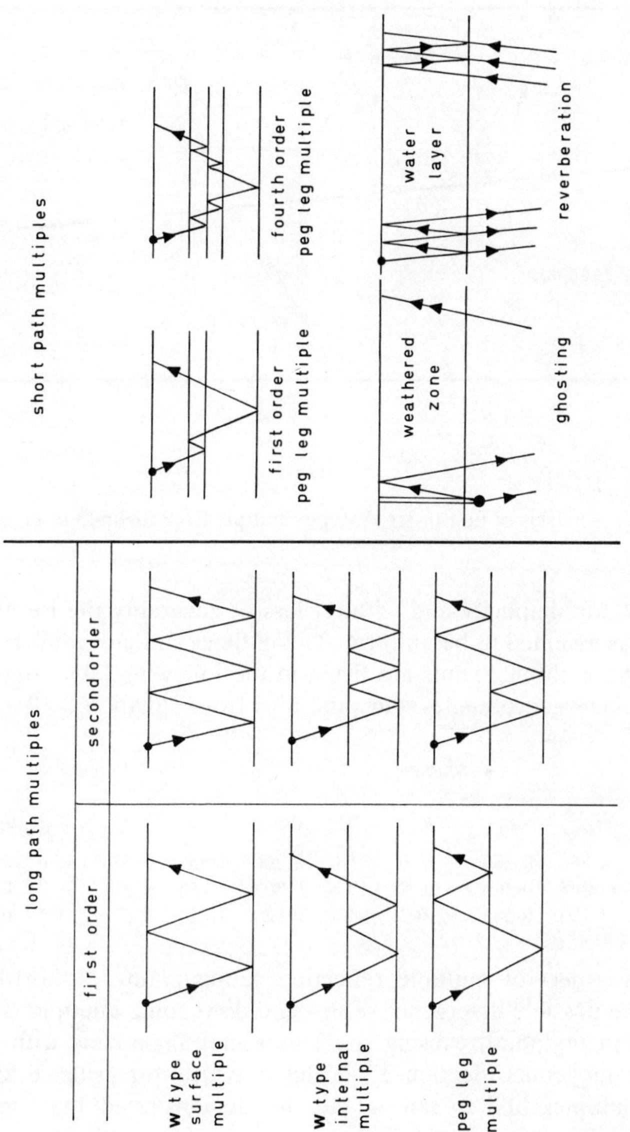

FIG. 3.32. Geometry and nomenclature multiple reflections.

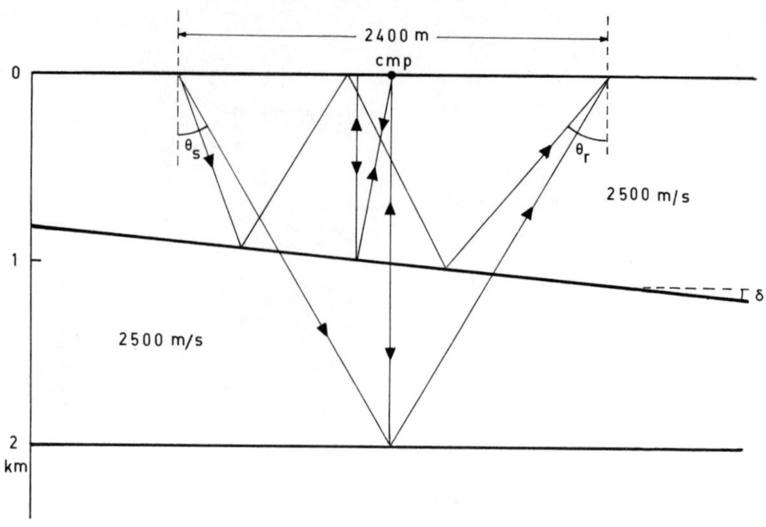

FIG. 3.33. Analysis of first order W-type multiple from inclined interface.

reflections; for simplicity and without loss of generality the formation velocity was assumed to be uniform. Travel times and moveouts for the primary and multiple events are listed in the following table, together with their emergence angles and time dips from equation 3.12:

		Multiple					*Primary*			
δ	t_0	t_{2400}	Δt	θ_s	θ_r	dt/dx	t_{2400}	Δt	$\lvert\theta_{r,s}\rvert$	dt/dx
0	1·600	1·866	0·266	−31·0	31·0	0·206	1·866	0·266	31·0	0·206
5	1·588	1·848	0·260	−20·8	40·8	0·202	1·856	0·286	31·2	0·207
10	1·552	1·795	0·243	−10·2	50·2	0·189	1·825	0·273	31·7	0·210

Another aspect of multiple reflection geometry in non-horizontal layers is the possible divergence of up- and downgoing multiple rays at the common midpoint, causing spurious signal alignments with non-hyperbolic moveouts (Section 3.7). This is typical for instance for all peg-leg multiples and it can in fact be demonstrated that, in the presence of dip, it is only the W-type multiples for which the up- and downgoing rays have the same angle of emergence at the c.m.p. In that respect it is instructive to study the peg-leg multiple geometry in Fig.

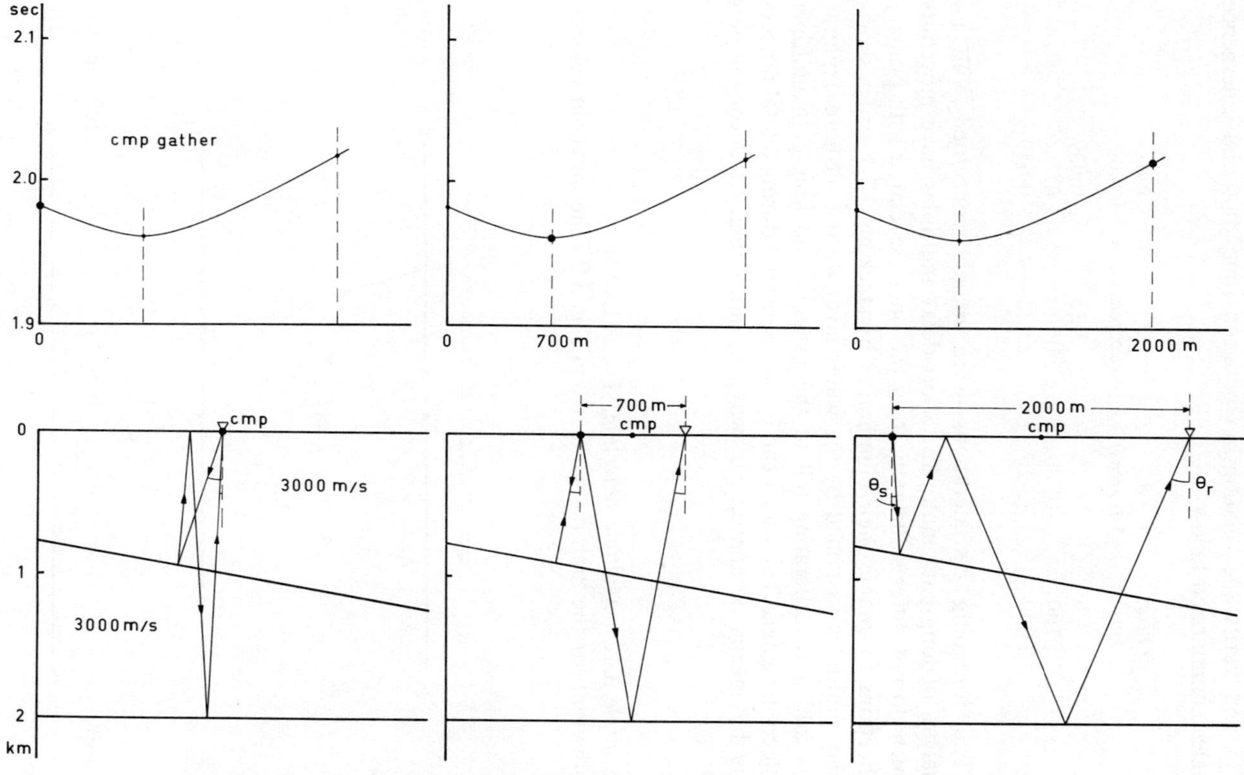

FIG. 3.34. Common midpoint geometry for peg-leg multiple from inclined interface.

3.34 and verify the relationship between time dips and **emergence** angles summarized below.

Offset (m)	θ_s (*degrees*)	θ_r (*degrees*)	dt/dx (s/km)
0	16·75	3·25	−0·039
700	10·00	10·00	0
2 000	−2·25	22·25	0·070

The foregoing considerations were mainly concerned with an analysis of long-path multiples. Short-delay multiples have moveouts which do not differ sufficiently from the ones associated with primary reflections to enable their elimination by stacking. High order short-delay multiples sustained by the water layer during offshore exploration add a reverberating tail to the primary and long-path multiple reflection signals. This effect can be compensated during digital processing by means of appropriate wavelet shortening, or deconvolution filters.

Example 3.12 *Multiple reflections I*
Show that for the multiple geometry in Fig. 3.35 the moveout velocity

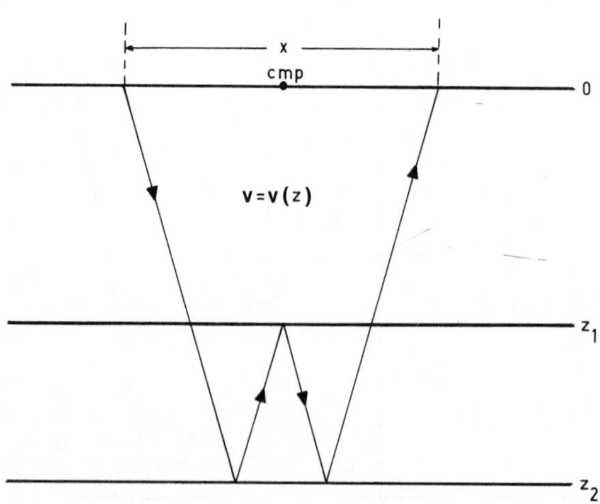

FIG. 3.35. Example 3.12.

is given by the relation

$$V_{mo}^2 = \frac{2\int_0^{z_2} V_z\,dz - \int_0^{z_1} V_z\,dz}{2\int_0^{z_2} \frac{dz}{V_z} - \int_0^{z_1} \frac{dz}{V_z}}$$

Example 3.12 *Solution*
From Section 3.4 we have

$$t_x = 4\int_0^{z_2} \frac{dz}{V_z(1-p^2V_z^2)^{\frac{1}{2}}} - 2\int_0^{z_1} \frac{dz}{V_z(1-p^2V_z^2)^{\frac{1}{2}}}$$

and

$$x = 4\int_0^{z_2} \frac{pV_z}{(1-p^2V_z^2)^{\frac{1}{2}}}\,dz - 2\int_0^{z_1} \frac{pV_z}{(1-p^2V_z^2)^{\frac{1}{2}}}\,dz$$

so that in consequence of Section 3.8,

$$V_{mo}^2 = \frac{1}{f^{(1)}(x^2)_0} = \frac{2\int_0^{z_2} V_z\,dz - \int_0^{z_1} V_z\,dz}{2\int_0^{z_2} \frac{dz}{V_z} - \int_0^{z_1} \frac{dz}{V_z}}$$

Example 3.13 *Multiple reflections II*
Determine in a horizontal velocity distribution $V_z = 1600 + 0{\cdot}6z$ m/s

1. The depth of a reflecting interface for which the normal incidence time of the primary reflection is equal to that of a first order surface multiple from a reflector at a depth of 1000 m.
2. The r.m.s. velocities at the levels of the two reflectors.
3. The moveouts of primary and multiple events for a shooting distance of 2400 m, estimated from the moveout velocities.

Example 3.13 *Solution*

1. From equation 3.10 the multiple time at the zero offset is $(4/0{\cdot}6)\ln(2200/1600) = 2{\cdot}123$ s.
   From equation 3.11 the depth of the second reflector is 2375 m.
2. Upon substitution of the linear velocity function in the appropriate expression for V_{rms} in Section 3.8, it obtains that the r.m.s.

velocities at 1000 and 2375 m are 1892 and 2275 m/s respectively.

3. From Example 3.12 the moveout velocity of the multiple is identical to the r.m.s. velocity at 1000 m. This leads to an estimated value for the moveout of the multiple of 0·350 s at the distance of 2400 m. The moveout of the primary is 0·248 s.

3.10 AREAL REFLECTION SURVEYS

Areal surveys are designed to provide multi-coverage reflection information along a dense grid of regularly spaced data points. Typical grid dimensions are 25×25 or 50×50 m, depending on reflector dip and signal frequencies. The principal objective of areal reflection surveys is the detailed mapping of hydrocarbon-bearing structures or their conjectured extensions by the three-dimensional migration of the stacked multi-coverage data (Chapter 8). Areal data acquisition on land is usually performed by shooting broadside into several recording spreads from a series of shot-point traverses. During marine work areal coverage may be obtained by in-line shooting along a dense system of parallel survey lines. After appropriate processing the migrated image of the subsurface can be displayed in the form of migrated time or true depth profiles in any direction across the target area, or as a sequence of horizontal slices exhibiting event outcrop configurations.

Fig. 3.36 is an example of part of a two-dimensional data acquisition scheme for land surveys. Its main elements comprise

1. Three north–south oriented recording traverses, lines A, B and C, spaced 900 m. Each traverse contains 33 geophone stations at 100 m intervals.
2. A square 100 by 100 m grid of shot points.

Multi-coverage reflection information is obtained by centre-spread broadside shooting into the three geophone traverses from east–west directed shot-point traverses, using 12-station recording spreads, as exemplified in the upper part of Fig. 3.36. This leads to a regular pattern of common midpoints of shot–station pairs, spaced 50 m in the north–south and east–west directions.

The multiplicity of the c.m.p. gathers can be expressed in terms of multiplicity components along two orthogonal directions. For the present case the coverage in the north–south direction is onefold in the

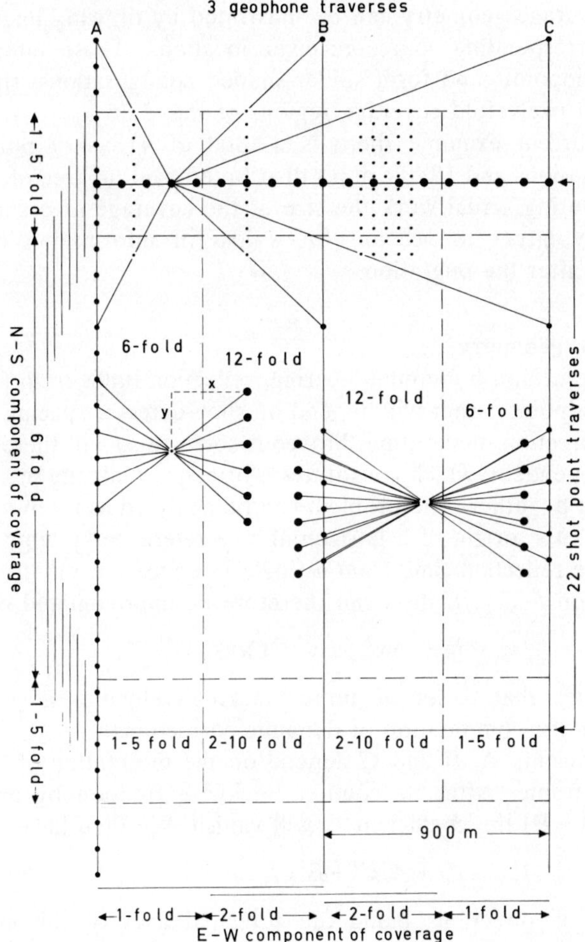

FIG. 3.36. Example of areal data acquisition scheme.

most northern and southern parts of the program and increases to sixfold in its central region. The east–west component of coverage varies from one to three, as can be inferred from reciprocity considerations by interchanging shot-point and geophone locations. In each part of the survey area the total multiplicity is the product of its north–south and east–west components.

Multi-coverage geometry can be illustrated by drawing lines which connect corresponding source–receiver locations. These intersect at common midpoints and form spider-shaped configurations, shown in the six- and twelvefold coverage regions of Fig. 3.36.

In the current example there is a total of 418 shot points, 99 recording stations and 1890 c.m.p.s distributed over an area of about 2 by 3 km. During actual work the size of the coverage area should be considerably larger to account for a loss of information near its boundaries after the migration.

Areal c.m.p. geometry

For the case of non-horizontal layering, reflection times from an areal common midpoint gather will depend on shot–detector spacing as well as on the direction of shooting. Moveout corrections will therefore be direction dependent, which constitutes a principal difficulty in stacking an areal c.m.p. gather. The problem can be analysed as follows. When the c.m.p. is the origin of a horizontal x–y reference system, we can write for the reflection time from a single interface $t(x, y) = t(-x, -y)$. The reflection t–x–y relation can therefore be approximated by

$$t_{xy}^2 = t_{00}^2 + Ax^2 + By^2 + Cxy$$

which implies that a set of three stacking velocities is generally required for the computation of dynamic corrections.

The coefficients A, B and C depend on the orientation of the x–y coordinate frame. After its counter-clockwise rotation by an angle $\frac{1}{2}\tan^{-1} C/(A-B)$ the coefficient C will vanish. We then have

$$t_{x'y'}^2 = t_{00}^2 + A'x'^2 + B'y'^2$$

the x' and y' directions being the principal axes of the moveout surface, corresponding to extreme values of stacking velocities. This can be illustrated by the special case of a uniform formation velocity for which, from Section 3.1

$$V_s = V(1 - \cos^2\omega \sin^2\delta)^{-\frac{1}{2}}$$

With the x'-axis pointing in the dip direction we have $\cos^2\omega = x'^2/(x'^2 + y'^2)$, so that

$$t_{x'y'}^2 = t_{00}^2 + (4/V^2)x'^2\cos^2\delta + (4/V^2)y'^2$$

with $A' \equiv (4/V^2)\cos^2\delta$ and $B' \equiv 4/V^2$.

The determination of coefficients A, B and C from a single c.m.p.

gather would require a highly complex velocity analysis procedure. Alternatively, t_{00}–A and t_{00}–B relations could separately be established at a limited number of grid points from in-line data in two orthogonal directions and estimated elsewhere by interpolation. Distributions for the coefficient C can then be obtained by a conventional one-dimensional velocity analysis of the areal gathers.

Ignoring the direction dependence of stacking velocities may lead to moveout correction errors which vary with shot–detector offset, reflector dip and recording time. These errors could be limited by reducing the spacing of recording traverses, narrowing the variance in direction of the c.m.p. spider legs.

REFERENCES

Al-Chalabi, M. (1979). Velocity determination from seismic reflection data. In: *Developments in Geophysical Exploration Methods I*, Applied Science, London.

Hubral, P. (1976). CDP ray modeling in the presence of 3-D plane isovelocity layers of varying dip and strike, *Geophysical Prospecting*, **24**(3).

Krey, Th. (1976). Computation of interval velocities from common reflection point move out times for *n* layers with arbitrary dips and curvatures in three dimensions when assuming small shot–geophone distances, *Geophysical Prospecting*, **24**(1).

Shah, P. M. (1973a). Ray tracing in three dimensions, *Geophysics*, **38**(3).

Shah, P. M. (1973b). Use of wave front curvature to relate seismic data with subsurface parameters, *Geophysics*, **38**(5).

Slotnick, M. M. (1936a). On seismic computations, with applications: I, *Geophysics*, **1**(1).

Slotnick, M. M. (1936b). On seismic computations, with applications: II, *Geophysics*, **1**(3).

BIBLIOGRAPHY

Slotnick, M. M. (1936a). On seismic computations, with applications: I, *Geophysics*, **1**(1).

Slotnick, M. M. (1936b). On seismic computations, with applications: II, *Geophysics*, **1**(3).

Green, C. H. (1938). Velocity determination by means of reflection profiles, *Geophysics*, **3**, 295–305.

Mott-Smith, M. (1942). Curved path methods applied to verticals and to wide shot spreads, *Geophysics*, **7**(2).

Deacons, L. E. (1943). An analysis of abnormal reflections, *Geophysics*, **8**, 1–13.

Krey, T. (1952). The significance of diffraction in the investigation of faults, *Geophysics*, **17**, 843.

Kaufman, H. (1953). Velocity functions in seismic prospecting, *Geophysics*, **18**(2).

Dix, C. H. (1955). Seismic velocities from surface measurements, *Geophysics*, **20**(1).

Kleyn, A. H. (1956). On seismic wave propagation in anisotropic media with applications in the Betun area, South Sumatra, *Geophysical Prospecting*, **4**(1).

Slotnick, M. M. (1959). *Lessons in Seismic Computing*, The Society of Exploration Geophysicists.

Mayne, W. H. (1962). Common reflection point horizontal data stacking techniques, *Geophysics*, **27**(6), part 2.

Grant, F. S. and West, G. F. (1965). *Interpretation Theory in Applied Geophysics*, McGraw-Hill, New York.

Courtier, W. H. and Mendenhall, H. L. (1967). Experience with multiple coverage seismic methods, *Geophysics*, **32,** 230–258.

Musgrave, A. W. (1967). *Seismic Refraction Prospecting*, The Society of Exploration Geophysicists.

Garotta, R. and Michon, D. (1967). Continuous analysis of the velocity function and of the move out corrections, *Geophysical Prospecting*, **15**(4).

Taner, M. T. and Koehler, F. (1969). Velocity spectra—Digital computer derivation and applications of velocity functions, *Geophysics*, **34**(6).

Cook, E. E. and Taner, M. T. (1969). Velocity spectra and their use in stratigraphic and lithologic differentiation, *Geophysical Prospecting*, **17**(4).

Trorey, A. W. (1970). A simple theory for seismic diffractions, *Geophysics*, **35**(5).

Robinson, J. C. (1970). Statistically optimal stacking of seismic data, *Geophysics*, **35**(3).

Hilterman, F. J. (1970). Three-dimensional seismic modelling, *Geophysics*, **35**(6).

Garotta, R. (1971). Selection of seismic picking based upon the dip, move out and amplitude of each event, *Geophysical Prospecting*, **19**(3).

Walton, G. G. (1971). Esso's 3-D seismic proves versatile, *Oil and Gas Journal*, **69** (March 29), 139–141.

Michon, D., Wlodarczak, R. and Merland, J. (1971). A new method of cancelling multiple reflections: 'Souston', *Geophysical Prospecting*, **19**(4).

Neidell, H. J. and Taner, M. T. (1971). Semblance and other coherency measures for multichannel data, *Geophysics*, **34,** 482–497.

Davis, J. M. (1972). Interpretation of velocity spectra through an adaptive modeling strategy, *Geophysics*, **37**(6).

Walton, G. G. (1972). Three dimensional seismic method, *Geophysics*, **37**(3).

Ocola, L. C. (1972). An algorithm for computation of source-to-detector ranges from the traveltime of multiply reflected sound waves, *Geophysics*, **37**(1).

Shah, P. M. (1973b). Use of wave front curvature to relate seismic data with subsurface parameters, *Geophysics*, **38**(5).

Shah, P. M. (1973a). Ray tracing in three dimensions, *Geophysics*, **38**(3).

Dunkin, J. W. and Levin, F. K. (1973). Effect of normal moveout on a seismic pulse, *Geophysics*, **38,** 635–642.

Krey, Th. (1976). Computation of interval velocities from common reflection point move out times for n layers with arbitrary dips and curvatures in three dimensions when assuming small shot–geophone distances, *Geophysical Prospecting*, **24**(1).

Hubral, P. (1976). CDP ray modeling in the presence of 3-D plane isovelocity layers of varying dip and strike, *Geophysical Prospecting*, **24**(3).

Berryhill, J. R. (1977). Diffraction response for nonzero separation of source and receiver, *Geophysics*, **42**(6).

Trorey, A. W. (1977). Diffractions for arbitrary source-receiver locations, *Geophysics*, **42**(6).

Levin, F. K. and Shah, P. M. (1977). Peg-leg multiples and dipping reflectors, *Geophysics*, **42**(5).

Ursin, B. (1977). Seismic velocity estimation, *Geophysical Prospecting*, **25**(4).

Levin, F. K. (1978). The reflection, refraction and diffraction of waves in media with an elliptical velocity dependence, *Geophysics*, **43**(3).

Tegland, E. R. (1978). Update concerning the three-dimensional seismic investigations in the Cottageville field, US Dep. Energy Spec. Publication no. MERC-77/5, 23–36.

May, B. T. and Straley, D. K. (1979). Higher order moveout spectra, *Geophysics*, **44**(7).

Kaila, K. L. and Krishna, V. G. (1979). A new computerized method for finding effective velocity from reversed reflection traveltime data, *Geophysics*, **44**(6).

Al-Chalabi, M. (1979). Velocity determination from seismic reflection data. In: *Developments in Geophysical Exploration Methods I*, Applied Science, London.

Brown, A. R. (1979). 3-D seismic survey gives better data, *Oil & Gas Journal* (Nov. 5), 57–71.

Douze, E. J. and Laster, S. J. (1979). Statistics of semblance, *Geophysics*, **44**(12).

Kennett, B. L. N. (1979). The suppression of surface multiples on seismic records, *Geophysical Prospecting*, **27**(3).

Johnson, J. P. and Bone, M. R. (1980). Understanding field development history using 3-D seismic survey, Annual AAPG-SEPM-EMO Convention Paper, Abstract *AAPG Bulletin*, **64**(5), 729.

Gardner, G. H. F. (1980). New dimensions in seismic exploration, Annual AAPG-SEPM-EMO Convention Paper, Abstract *AAPG Bulletin*, **64**(5), 711.

Blackburn, G. (1980). Errors in stacking velocity—True velocity conversion over complex geologic situations, *Geophysics*, **45**(10).

Grimlin, D. R. and Smith, J. W. (1980). A comparison of seismic trace summing techniques, *Geophysics*, **45**(6).

Lash, C. C. (1980). Shear waves, multiple reflections, and converted waves found by a deep vertical wave test (vertical seismic profiling), *Geophysics*, **45**(9).

Hubral, P. (1980). Computation of the normal movement velocity in 3-D laterally inhomogeneous media with curved interfaces, *Geophysical Prospecting*, **28**(2).

Hubral, P. and Krey, Th. (1980). *Interval Velocities from Seismic Reflection Time Measurements*, K. L. Larner, Western Geophysical Company, Houston, Texas.

Khattri, K., Vig, P. K. and Rao, N. D. J. (1980). An optimized technique for determining stacking velocity from seismic reflection data, *Geophysical Prospecting*, **28**(6).

Rietsch, E. (1980). Estimation of the signal-to-noise ratio of seismic data with an application to stacking, *Geophysical Prospecting*, **28**(4).

Brown, A. R. and MacBeath, R. G. (1980). Three-D seismic surveying for field development comes of age, *Oil & Gas Journal* (Nov. 17).

Brown, A. R., Dahm, C. G. and Graebner, R. J. (1981). A stratigraphic case history using three-dimensional seismic data in the Gulf of Thailand, *Geophysical Prospecting*, **29**(3).

Greenhalgh, S. A. and King, D. W. (1981). Curved raypath interpretation of seismic refraction data, *Geophysical Prospecting*, **29**(6).

Bruland, L. (1981). Velocity analysis using iterative stackings, *Geophysical Prospecting*, **29**(1).

Hajnal, Z. and Sereda, I. T. (1981). Maximum uncertainty of interval velocity estimates, *Geophysics*, **46** (November).

Stoffa, P. L., Diebold, J. B. and Buhl, P. (1982). Velocity analysis for wide aperture seismic data, *Geophysical Prospecting*, **30**(1).

CHAPTER 4

Static Corrections

4.1 PRELIMINARY REMARKS

Static corrections applied to seismic field data compensate for travel time differences due to elevation changes and to lateral fluctuations of the weathering velocity and the weathering thickness. The weathered layer is also referred to as the unconsolidated layer or the low velocity layer. Its average compressional wave velocity is of the order of 600 m/s, whereas that of the consolidated layer underneath it normally varies from about 1600 to 2000 m/s.

The effect of applying weathering corrections is a replacement of the weathered layer by the material of the consolidated formation just below it. Subsequent elevation corrections move both sources and recording stations down to a common reference plane, the seismic datum plane. The latter is usually identified on reflection profiles with respect to mean sea level. The total static correction is the sum of weathering and elevation corrections.

Provisional static corrections determined in the field represent the 'field statics'. During reflection work their computation is based on the simplifying assumption of vertical emergence of reflected rays and on a single layer weathering model whose velocity is uniform along a reflection spread. In most areas, however, the base of the weathering corresponds to a transition zone between unconsolidated and consolidated material. When the precision of the field statics is not sufficient to produce a satisfactory stack, residual static corrections may be obtained from auxiliary static correction programs. These are based on cross-correlations of moveout corrected traces on c.m.p. gathers to which the provisional static corrections have already been applied.

4.2 WEATHERING CORRECTIONS

Of fundamental importance in the theory of static corrections is a practical definition of the weathering correction. From the previous section it is for the case of vertical emergence given by the relation

$$C_w = \frac{z}{V_w} - \frac{z}{V_c}$$

where z is the weathering thickness and V_w and V_c the velocities of the weathered and consolidated layers.

Since the weathered layer is relatively thin, its velocity and thickness cannot be determined from conventional refraction data. Another elusive quantity is the vertical weathering time z/V_w; it is not identical to the weathering correction, but can at shot-point locations be related to the uphole time t_u by

$$t_u = \frac{d_s - z}{V_c} + \frac{z}{V_w} \qquad \text{for } d_s \geqslant z$$

where d_s is the shot depth. From this relation we have

$$C_w = t_u - d_s/V_c$$

For any position of the seismic source in the weathered interval approximate values for C_w can be derived from delay times in the weathered layer provided by refraction data on reflection field records or by separate short-distance refraction profiles. The relevant near-surface refraction geometry is illustrated in Fig. 4.1. If z_1 is the weathering thickness below the source and z_2 that below a recording station, it follows from wavefront properties that

$$t(x) = \frac{x}{V_c} + \frac{(z_1 + z_2)\cos\theta_w}{V_w}$$

Presuming that V_c is known, the delay time is a measurable quantity given by

$$D_w = \frac{(z_1 + z_2)\cos\theta_w}{V_w} = \frac{(z_1 + z_2)(1 - V_w^2/V_c^2)^{\frac{1}{2}}}{V_w}$$

from which after some algebra

$$C_w = D_w \frac{(V_c - V_w)^{\frac{1}{2}}}{(V_c + V_w)^{\frac{1}{2}}} = D_w K$$

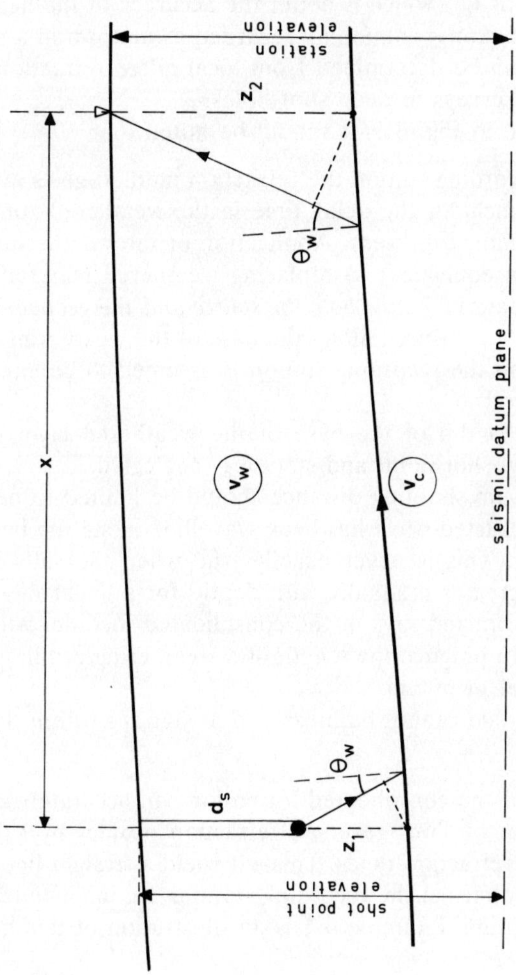

FIG. 4.1. Near-surface refraction geometry.

K is a conversion factor; it is smaller than one and rather insensitive to moderate variations of V_c and V_w. When for instance $V_w = 600$ m/s and $V_c = 1700$ m/s, the value of K is 0·692. After increasing V_w to 750 m/s, K becomes 0·623. This corresponds to an error of 10% in the estimated value of C_w, which is about the accuracy of the field statics. The conversion factor is commonly regarded as uniform in a restricted area; its value can be determined from local micro refraction spreads or from uphole surveys in deep shot holes.

With reference to Fig. 4.1, it should be noted that

1. At each recording station the refraction method gives a value for D_w which includes the delay time in the weathered zone underneath the shot, if present. Application of the weathering correction is then equivalent to replacing weathered material by consolidated material below *both the source and the geophone station.*
2. The distance measured along the base of the weathering between the shot and the recording station is assumed to be equal to the horizontal offset x.
3. The effect of dip of the base of the weathered layer on delay times below shot point and station is neglected.
4. The maximum shooting distance should be limited to be assured that the refracted wave has been travelling along the base of the weathering. This is never exactly true when the subweathering velocity increases gradually with depth; for a linear increase, for instance, refracted rays in the consolidated medium will be arcs of circles. In practice the length of a weathering profile is usually of the order of 600 m.
5. The delay time cannot be measured at stations within the critical distance.

The velocity of the consolidated formation can accurately be determined from a set of short reversed refraction profiles by subtracting common station refraction times. This will yield a straight line of slope $2/V_c$ since delay times at the recording stations are independent of the direction of shooting. Example 4.1 is an illustration of this method.

4.3 ELEVATION CORRECTIONS

The total elevation correction for a specific shot–station pair is defined by

$$C_e = \frac{(e_s - d_s) + e_g}{V_c}$$

where e_s and e_g are shot-point and geophone station elevations with reference to the seismic datum plane, and d_s is the shot depth. The term $(e_s - d_s)/V_c$ is referred to as the shot hole correction (s.h.c.). The seismic datum plane should always be near the ground surface to avoid the effect of an increase of the subweathering velocity on elevation corrections. An average value of V_c based on refraction observations can then be used as a uniform elevation correction velocity. In lowlands the seismic datum plane is usually at mean sea level, whereas in elevated regions it may be several hundred metres higher.

4.4 STATIC CORRECTIONS FROM UPHOLE TIMES

This procedure requires that seismic sources are placed at or below the base of the weathering. Recalling that $C_w = t_u - d_s/V_c$, the static correction at a shot point is given by

$$C_g = t_u + (e_s - d_s)/V_c = t_u + \text{s.h.c.}$$

C_g is thus the sum of elevation and weathering corrections for a *fictive geophone station* at the shot point.

The uphole method is widely used during multi-coverage data acquisition, where near-surface conditions are sufficiently uniform to warrant the interpolation of static corrections derived from uphole times. For a specific reflection trace the final static correction is then the sum of an interpolated value of C_g and the elevation correction at some associated shot point. A related but more extensive procedure is based on the picking of reflection times from a shallow horizon; this will further be illustrated in Example 4.2.

4.5 OFFSHORE STATICS

During marine surveys source and streamer depth corrections are comparable to elevation corrections on land. For source and streamer positions of 8 and 10 m below mean sea level for instance and a water velocity of 1500 m/s, the total static correction would be −12 ms. The effect of vertical tidal movements on the position of the recording system with respect to mean sea level may lead to static time shifts of the order of 4 ms, which is usually ignored, however.

When there is a significant water/sea-bed velocity contrast a special static correction may be required to account for the influence of sea bottom topography on reflection times. In analogy with the weathering correction, a water delay correction can then be defined by

$$C_{wtr} = (z_1 + z_2)\left(\frac{1}{1500} - \frac{1}{V_{sb}}\right)$$

where V_{sb} is the velocity of the sea bottom material, and z_1 and z_2 are the sea bottom depths at source and recording station with reference to a convenient datum plane.

Example 4.1 *Static corrections I*

For a reversed refraction profile—spread length 500 m, station spacing 50 m—first arrival times and station elevations are listed in the following table:

Station	*Shot* S_1 *east* (ms)	*Shot* S_2 *west* (ms)	*Elevation* (m)
1	71	405	10
2	110	382	12
3	129	336	15
4	161	300	16
5	196	277	16
6	232	253	17
7	259	218	20
8	285	180	22
9	301	135	24
10	342	112	22
11	374	82	20

Shot point S_1: 50 m west of station 1, elevation 11 m, shot depth 18 m.
Shot point S_2: 50 m east of station 11, elevation 20 m, shot depth 12 m.

Determine

1. The subweathering velocity.
2. Weathering corrections for shots S_1 east and S_2 west.
3. Elevation corrections at recording stations.
4. The total static corrections for reflection recordings from shot points S_1 and S_2

Conversion factor 0·7. Scales: 1 mm = 2 m, 1 mm = 2 ms.

Example 4.1 *Solution*

The subweathering velocity is 1600 m/s. Its graphical determination with the subtraction method is illustrated in Fig. 4.2. The static

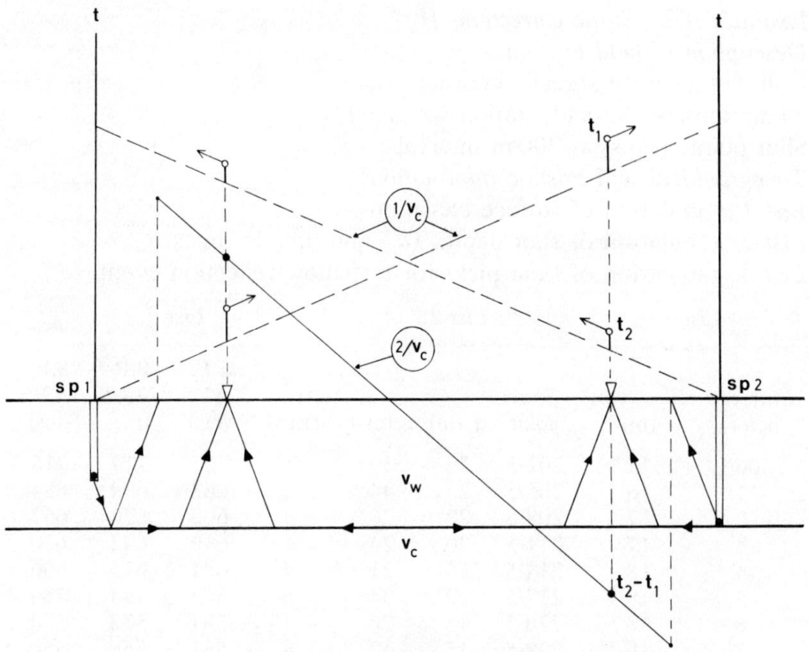

FIG. 4.2. Example 4.1 solution.

correction computations are summarized in the following table:

Stn	*D_w measured* (ms) S_1	S_2	*Diff.*	*D_w adjusted for constant diff.* (ms) S_1	S_2	*C_w* S_1	S_2	*Elevation corr.* (ms)	*Total statics* S_1 −4	S_2 +5
1	40	61	21	40	61	28	43	6	30	54
2	48	70	22	49	69	34	48	7	37	60
3	35	55	20	35	55	24	38	9	29	52
4	36	50	14	33	53	23	37	10	29	52
5	40	58	18	39	59	27	41	10	33	56
6	45	66	21	45	66	31	46	11	38	62
7	40	62	22	41	61	29	43	12	37	60
8	35	55	20	35	55	24	38	14	34	57
9	20	41	21	20	41	14	29	15	25	49
10	30	50	20	30	50	21	35	14	31	54
11	30	51	21	30	51	21	36	12	29	53
			20 Average							

Example 4.2 *Static corrections II*
Description of field technique:
Roll along centre spread shooting.
24 stations per spread, station spacing 100 m.
Shot points (s.p.s) at 300 m intervals.
Topographical and seismic information:
List 1: tabulation of surface elevations.
List 2: tabulation of shot depths (d_s) and uphole times (t_u).
List 3: tabulation of time picks for a shallow reflection event.

List 1

Surface point	*Elevation* (m)
200	15
1	16
2	17
3	17
4	18
5	18
6	18
7	19
8	20
9	21
210	21
1	20
2	21
3	19
4	21
5	23
6	24
7	25
8	26
9	26
220	25
1	24
2	23
3	23
4	22
5	22
6	22
7	23
8	23
9	23
230	23

List 2

Shot point	d_s (m)	t_u (ms)
202·5	20	39
205·5	25	44
208·5	20	26
211·5	20	24
214·5	15	21
217·5	20	34
220·5	15	26
223·5	15	17
226·5	10	10

List 3

Trace	S.P. 211·5 (ms)	S.P. 214·5 (ms)	S.P. 217·5 (ms)
1	707	717	715
2	683	694	684
3	660	670	657
4	640	642	630
5	621	615	606
6	603	594	586
7	581	573	570
8	561	556	550
9	548	544	544
10	535	536	540
11	527	525	536
12	524	528	540
13	525	533	539
14	522	537	539
15	533	549	543
16	547	557	546
17	560	566	555
18	581	580	567
19	597	590	583
20	614	607	599
21	634	626	621
22	651	648	643
23	674	670	668
24	697	696	694

Trace numbers increase in the same order as surface point numbers; on the record from s.p. 211·5 trace 1 corresponds to surface point 200. At all shot points the seismic source is below the base of the weathering. Subweathering velocity is 1700 m/s.

1. Determine the shot hole correction (s.h.c.) and the static correction (C_g) at each shot point.
2. Derive graphically the station corrections (sum of weathering and elevation corrections) at surface points 209 to 220 from the reflection time data in list 3. Hints: estimate reflection times at shot points by interpolation between adjacent traces and apply the static corrections C_g. Measure deviations of recorded times from a smooth hyperbolic curve drawn through the control points. Exploiting the multiplicity of the information the final estimate of a static correction value will be the average of three measurements.
3. Tabulate the total static corrections at surface points 209 to 220 for recordings from shot points 211·5 and 217·5.

Recommended scales: 1 cm = 100 m, 1 mm = 2 ms.

Example 4.2 *Solution*

1. Shot hole corrections (s.h.c.) and static corrections (C_g):

Shot point	e_s (m)	d_s (m)	s.h.c. = $(e_s - d_s)/1700$ (ms)	*Static correction* t_u + s.h.c. (ms)
202·5	17	20	−2	37
205·5	18	25	−4	40
208·5	20·5	20	0	26
211·5	20·5	20	0	24
214·5	22	15	4	25
217·5	25·5	20	3	37
220·5	24·5	15	6	32
223·5	22·5	15	4	21
226·5	22·5	10	7	17

2. From Fig. 4.3, station corrections can be tabulated as follows:

Surface point	*Stn corr. shot* 211·5	*Stn corr. shot* 214·5	*Stn corr. shot* 217·5	*Average of three*
209	24	24	25	24
210	23	22	24	23
211	23	22	23	23

Surface point	*Stn corr. shot* 211·5	*Stn corr. shot* 214·5	*Stn corr. shot* 217·5	*Average of three*
212	24	22	24	23
213	18	17	18	18
214	22	22	23	22
215	28	28	28	28
216	29	29	30	29
217	37	37	38	37
218	37	36	37	37
219	34	33	35	34
220	34	33	34	34

3. Total static corrections:

Surface point	*Total st. corr. for record from s.p.* 211·5 (+0)	*Total st. corr. for record from s.p.* 217·5 (+3)
209	24	27
210	23	26
211	23	26
212	23	26
213	18	21
214	22	25
215	28	31
216	29	32
217	37	40
218	37	40
219	34	37
220	34	37

Example 4.3 *Oblique emergence*

Derive expressions for the weathering and elevation corrections as a function of the direction angle of the reflected raypath in the consolidated formation just below the base of the weathering. Seismic datum plane at base weathering.

Example 4.3 *Solution*

1. From Fig. 4.4(a), subtracting the time spent in the weathered layer will contribute $z/(V_w \cos\alpha)$ to the weathering correction. The replacement of V_w by V_c material is equivalent to addition of the travel time along ray segment AB. Therefore

$$C_w = \frac{z}{V_w \cos\alpha} - \frac{z}{V_c \cos\alpha}\cos(\theta-\alpha)$$

with $\sin\alpha/\sin\theta = V_w/V_c$.
For $\theta = 0°$, $C_w = z/V_w - z/V_c$ (vertical emergence).
For $\theta = 90°$, $C_w = D_w$ (refracted ray).

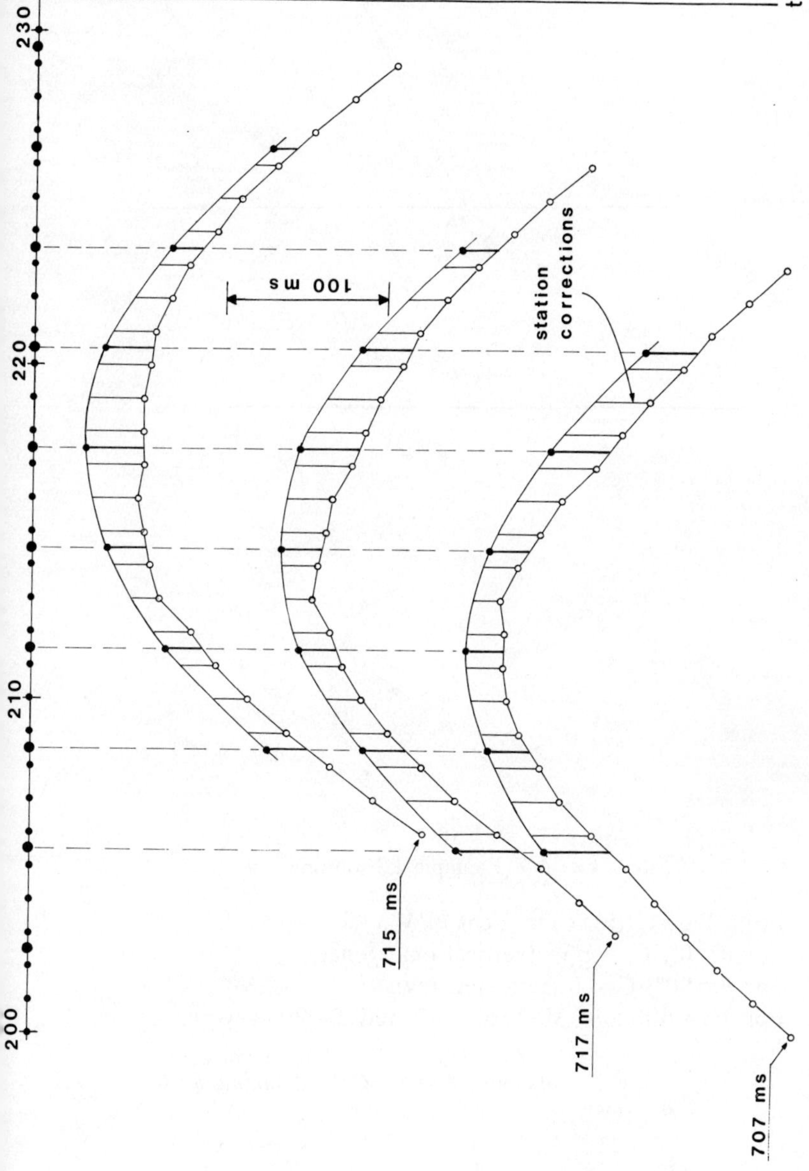

FIG. 4.3. Example 4.2 solution.

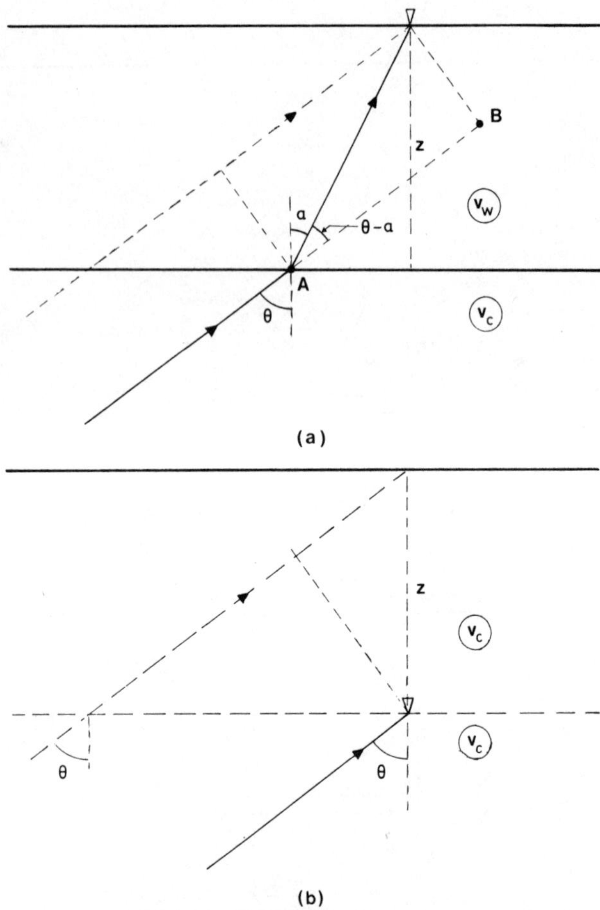

FIG. 4.4. Example 4.3 solution.

2. From Fig. 4.4(b), $C_e = (z \cos \theta)/V_c$.
   For $\theta = 0°$, $C_e = z/V_c$ (vertical emergence).
   For $\theta = 90°$, $C_e = 0$ (refracted ray).
   For $V_w = 600$ m/s, $V_c = 1600$ m/s and $z = 20$ m we have:

θ	C_w (ms)	*Error* C_w *assuming* $\theta = 0$ (ms)	C_e (ms)	*Error* C_e *assuming* $\theta = 0$ (ms)
0°	21	0	12	0
30	22	1	11	−1
60	25	4	6	−6
90	31	10	0	−12

BIBLIOGRAPHY

Hagedoorn, J. G. (1959). The plus–minus method of interpreting seismic refraction sections, *Geophysical Prospecting*, **7**(2).

Hileman, J. A., Embree, P. and Pflueger, J. C. (1968). Automatic static corrections, *Geophysical Prospecting*, **16**(3).

Laski, J. D. (1970). Simultaneous estimation of parameters of reflection events (depth, dip, velocity) and relative static corrections, *Geophysical Prospecting*, **18**(2).

Disher, D.A. and Naquin, P. G. (1970). Statistical automatic statics analysis, *Geophysics*, **35**(4).

Gendzwill, D. J. (1978). A method of weathering correction, *Geophysical Prospecting*, **26**(3).

Larner, K. L. Gibson, B., Chambers, R. and Wiggins, R. A. (1979). Simultaneous estimation of residual statics and crossdip time corrections, *Geophysics*, **44**(7).

Rogers, A. W. (1979). Determination of static corrections. In: *Developments in Geophysical Exploration Methods II*, Applied Science, London.

Zelei, A. and Saghy, G. (1980). Residual static corrections—Iterative solution analysis, *Geophysical Prospecting*, **28**(2).

CHAPTER 5

Reflection, Transmission and Acoustic Impedance

5.1 BASIC CONCEPTS

The reflection process

In seismic work the reflection coefficient is defined as the ratio of the amplitudes of incident and reflected waves. For a given lithologic contrast it depends on the angle of incidence and on incident and reflected wave types. In practice the effect of oblique incidence is usually neglected, however, and the reflection coefficient expressed as a simple function of the densities and compressional wave velocities in incident and transmitting media.

The reflection coefficient may be positive or negative, a negative value corresponding to a 180° phase shift of the frequency components of an incident wavelet upon reflection. We shall analyse this aspect of polarity inversion for the normal incidence case with reference to Fig. 5.1. The waveforms shown are graphical analogues of earth particle displacements induced by a compressional wave. A displacement vector in the region of the front of the wavelet and which points in the direction of propagation indicates a compression, otherwise the front of the wavelet is a rarefaction. Boundary conditions require that during reflection the motions of earth particles just above and below the reflecting interface are identical. This implies that the sum of incident and reflected displacement vectors must be equal to the transmitted displacement vector. When there is no phase inversion the front of the reflected wave will be a compression again. From Fig. 5.1(a) the reflected displacement vector of magnitude RA_i is then pointing upward, A_i being the amplitude of the incident wave and R the reflection coefficient. Consequently, the value of the transmitted displacement vector is $(1-R)A_i$. The factor $(1-R)$ is the transmission coefficient T, the ratio of transmitted to incident displacement amplitudes. When phase inversion occurs [Fig. 5.1(b)], the front of the reflected wave becomes a rarefaction and the reflected displacement vector is pointing downward. Its magnitude is now $-RA_i$, which is positive since R is negative. The transmission coefficient is thus $(1-R)$, as before. A reflection accompanied by phase inversion is often referred to as a 'soft kick'; the transmission coefficient is then larger than one. Otherwise the reflection is a 'hard kick' and the transmission coefficient is between zero and one. We further have that during reflection pressures at contiguous points on opposite sides of the reflecting interface should be equal. Since in the incident medium the pressure just above the reflector is proportional to $(1+R)A_i$, the transmission coefficient for pressure amplitudes will be $(1+R)$.

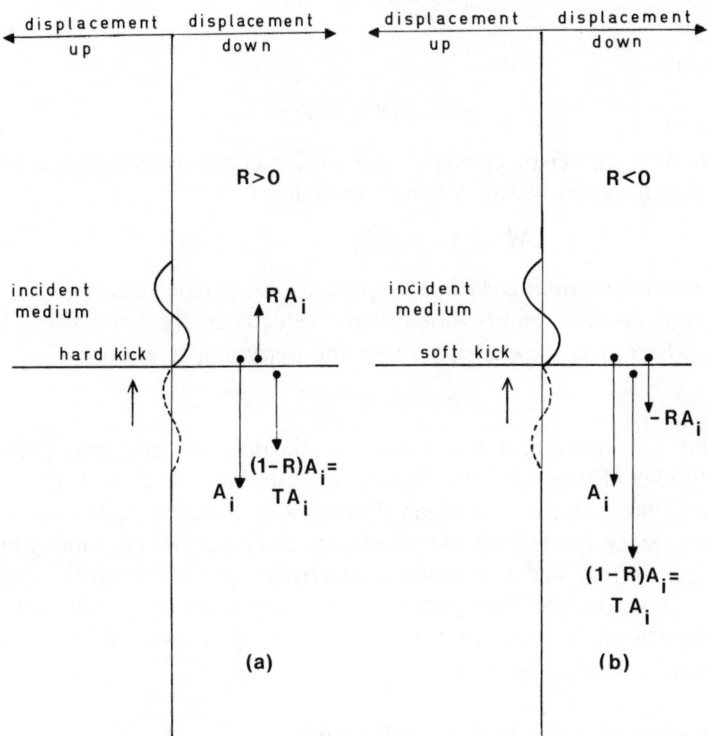

FIG. 5.1. Reflection and transmission processes for positive and negative reflection coefficients.

Acoustic impedance

Depending on the physical quantity of interest a plane compressional wave moving in the direction of the positive z-axis with a velocity V can be represented by any of the following relations:

Particle displacement:

$$u = S(z - Vt)$$

Particle velocity:

$$q = \frac{\partial u}{\partial t} = -VS'(z - Vt)$$

Specific strain:

$$s = \frac{\partial u}{\partial z} = S'(z - Vt)$$

Pressure:

$$p = -MS'(z - Vt)$$

where M is an elastic constant, defined for non-porous media in terms of Poisson's ratio σ and Young's modulus E by

$$M = [(1 - \sigma)E]/[(1 + \sigma)(1 - 2\sigma)]$$

The elasticity modulus M relates pressure to specific strain by $p = -Ms$ and controls the compressional wave velocity in the expression $V^2 = M/\rho$, where ρ is the density. From the foregoing relations

$$p/q = P/Q = M/V = \rho V$$

P and Q represent r.m.s. values of particle velocity and pressure amplitudes. From acoustic theory the product of P and Q is the acoustic intensity, or the mean flow of seismic energy per unit cross-section. Since $P = (\rho V)Q$, the quantities P, Q and ρV are analogous to voltage, current and resistance in electricity. ρV is therefore referred to as the acoustic impedance. The dimensionless specific acoustic impedance of a rock unit is defined as the ratio of its acoustic impedance to that of water.

Evaluation of the reflection coefficient

For the normal incidence case an incident flow of compressional wave energy will be equal to the sum of reflected and transmitted energy flows. Hence

$$\rho_1 V_1 Q_1^2 = \rho_1 V_1 R^2 Q_1^2 + \rho_2 V_2 (1-R)^2 Q_1^2$$

or

$$\rho_1 V_1 = \rho_1 V_1 R^2 + \rho_2 V_2 (1-R)^2$$

the subscripts 1 and 2 referring to the incident and transmitting media respectively. After solving the latter equation we obtain for the reflection coefficient

$$R = \frac{\rho_2 V_2 - \rho_1 V_1}{\rho_2 V_2 + \rho_1 V_1}$$

Remark: Pressures in incident and transmitting media are proportional to $\rho_1 V_1$ and $\rho_2 V_2(1-R)$ respectively. The transmission coefficient for pressure amplitudes can therefore be written in the form $\rho_2 V_2(1-R)/\rho_1 V_1$, which simplifies to $1+R$ upon substitution of the above expression for R, as can be expected.

Logarithmic expression of *R*

When for two layers in contact $\Delta(\rho V)$ denotes the difference between their acoustic impedances and $(\rho V)_{av}$ the average value, we can write for the reflection coefficient

$$R = \frac{\Delta(\rho V)}{2(\rho V)_{av}}$$

which for small acoustic impedance contrasts can be approximated by

$$R \simeq \frac{\ln(\rho V)}{2}$$

in view of

$$\frac{d \ln(\rho V)}{d(\rho V)} = \frac{1}{\rho V}$$

The reflection coefficient is thus nearly equal to half the difference between the natural logarithms of two acoustic impedances. When for instance the densities in incident and transmitting media are 2·1 and 2·2, and the corresponding velocities 2000 and 2200, the true value for R is 0·0708 and its approximation from the logarithmic expression 0·0709.

Reflection coefficient and acoustic impedance logs

These are graphical displays of R and $\ln(\rho V)$ as a function of depth or two-way vertical time. Their mutual relationship is illustrated in Fig.

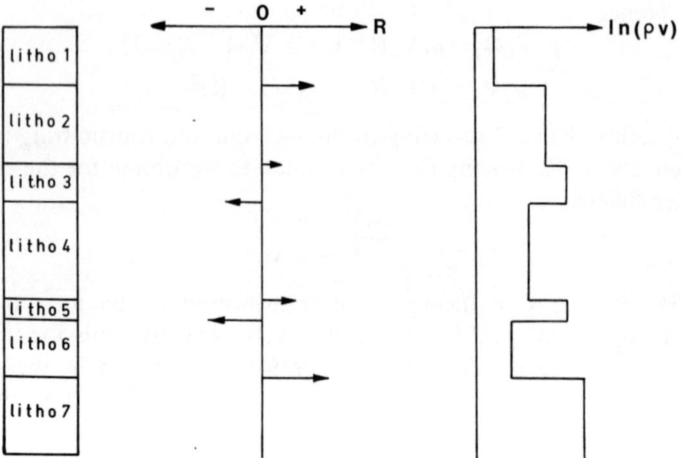

FIG. 5.2. Relationship between reflection coefficient and acoustic impedance logs.

5.2, from which we may summarize that

1. The reflection coefficient log is a record of the differences between successive logarithmic acoustic impedances.
2. An acoustic impedance log is the running sum of a sequence of reflection coefficients.

Fig. 5.2 shows a blocked acoustic impedance function. When the acoustic impedance changes gradually the reflection response of the earth can be determined through a more general analysis based on linear filter theory. This will be discussed in Section 5.3.

Two-way transmission

For each acoustic contrast the sign of the reflection coefficient depends on which side of it is the incident medium. When for a wave travelling in the downward direction the value of the reflection coefficient at a layer interface is r, it will be $-r$ for an upgoing wave. The corresponding transmission coefficients are then $1-r$ and $1+r$, and the two-way transmission factor is $1-r^2$ for pressure as well as displacement waves. For a sequence of n layer interfaces with reflection coefficients $r_1, r_2, \ldots, r_n$, the amplitude of a reflection from the nth interface would thus be proportional to

$$(1-r_1^2)(1-r_2^2)\ldots(1-r_{n-1}^2)r_n$$

after correcting for the effect of spherical divergence and absorption. With a few exceptions reflection coefficients in sedimentary sections are of the order of 0·1, which implies that for a single interface the two-way transmission factor is usually nearly equal to one. Similarly, for a limited number of moderate acoustic contrasts associated with a sequence of relatively thick layers the effect of transmission coefficients on signal amplitude will be rather small. In thinly bedded sedimentary sections, however, there are hundreds of reflecting interfaces. For 200 of these the product of all two-way transmission factors will be of the order of $0{\cdot}99^{200}$ or 0·13 and for 300 interfaces about 0·05 etc. Such values are contrary to observations.

The above analysis does not take account, however, of the effect of short-delay peg-leg multiples on reflection amplitudes. Due to the cyclic nature of a deposition process there is a general tendency that the sign of reflection coefficients in sedimentary sections will change persistently. Under such a condition a large number of short-path peg-leg multiples will interfere constructively with primary events. This is illustrated in Fig. 5.3 for an alternation of two lithologies. In each layer reflection coefficients for internal reflections have the same polarity. Short-delay peg-leg multiples in the wake of a primary or long-path multiple reflection are therefore nearly in phase with it and interfere constructively. The resultant reinforcement of primary and multiple events will be more effective for the lower frequencies. Interference by short-path multiples during transmission is thus a frequency dependent process, a property in common with absorption.

'True amplitudes' of reflection signals are proportional to reflection coefficients. During true amplitude processing all of the above effects, inclusive of that of absorption, on reflection amplitudes is usually described by a single relation of the form $A(t) = A(0)e^{-\alpha t}$, where e is the base of natural logarithms and t the recording time. The factor α is an empirical quantity estimated from the trend of amplitude variations after gain recovery and correction for geometric spreading. A typical value for α is 0·7, which corresponds to an amplitude decline of 6 dB/s. These correction procedures are not very exact and will produce reflection traces with amplitudes which are only approximately proportional to reflection coefficients.

Example 5.1 *Reflection amplitudes*

The formation below a seismic source at the ground surface consists of alternating sands and shales with specific acoustic impedances of 2·65

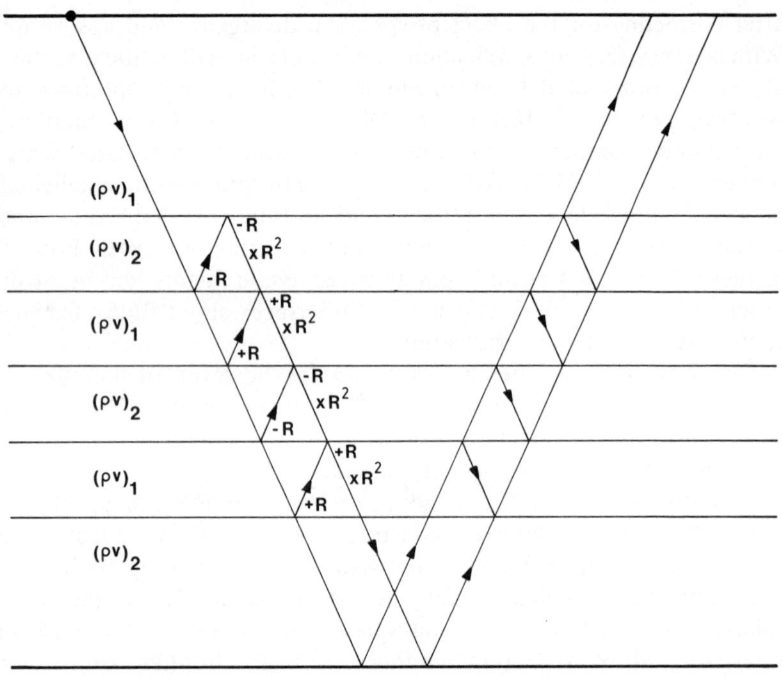

FIG. 5.3. Reinforcement of primary reflection by short-delay peg-leg multiples.

and 2·35 respectively. The layer thickness is 10 m. Neglecting absorption, determine the ratio of amplitudes of normal incidence reflections from interfaces at depths of 500 and 1000 m when first order peg-leg multiples in the layers interfere constructively with primary events. Evaluate spherical divergence for a uniform velocity.

Example 5.1 *Solution*

Effect of acoustic impedance contrasts: The amplitude of a primary reflection from the bottom of the nth layer is $C(1-r^2)^{n-1}r$, whereas that of the integrated peg-leg multiples is $C(1-r^2)^{n-1}nr^3$. After adding we get for the reflection amplitude $C(1-r^2)^{n-1}(1+nr^2)r$.

Upon substitution of $r=0{\cdot}06$ and the appropriate values for n (50 and 100), the amplitude ratio A_{1000}/A_{500} becomes 0·96.

Effect of spherical divergence: Due to spherical divergence reflection amplitudes decrease inversely proportional to the distance travelled. In

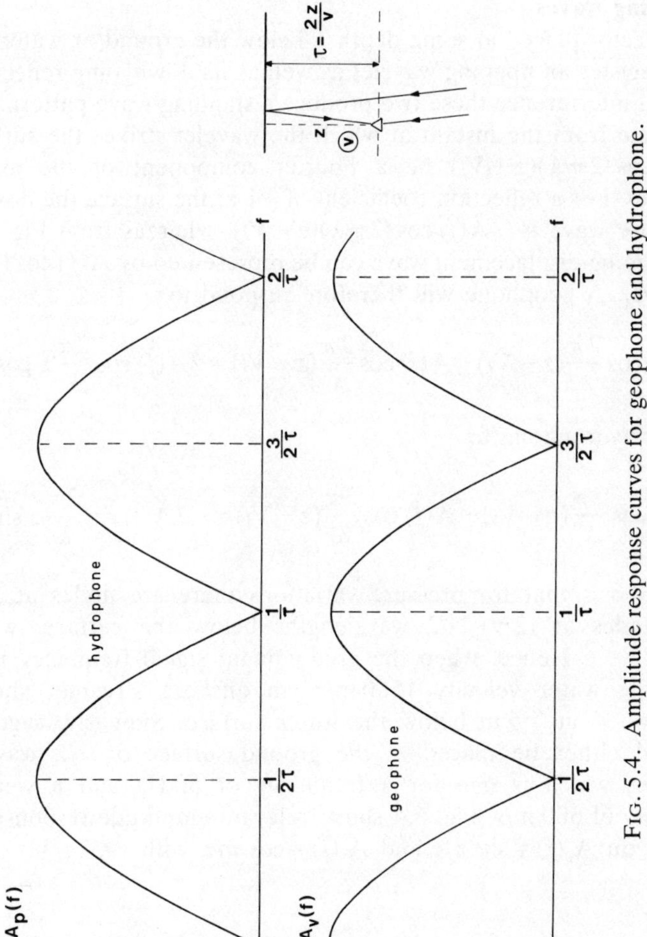

Fig. 5.4. Amplitude response curves for geophone and hydrophone.

the present case the amplitude ratio is thus reduced by a factor 0·5. Its final value is then $0{\cdot}96\times 0{\cdot}5 = 0{\cdot}48$, neglecting the small difference in lengths of primary and multiple paths.

Standing waves

A detector placed at some depth z below the ground or water surface will register an upgoing wavelet as well as its downgoing reflection. By spatial interference these two produce a standing wave pattern. Counting time from the instant at which the wavelet strikes the surface, let $A(f)\cos(2\pi/\lambda)(z+Vt)$ be a Fourier component of the ascending wavelet. For a reflection coefficient of -1 at the surface the downgoing pressure wave is $-A(f)\cos(2\pi/\lambda)(z-Vt)$, whereas from Fig. 5.1 the downgoing displacement wave can be represented by $A(f)\cos(2\pi/\lambda)\times(z-Vt)$. A geophone will therefore respond to

$$A(f)\cos\frac{2\pi}{\lambda}(z+Vt)+A(f)\cos\frac{2\pi}{\lambda}(z-Vt)=2A(f)\cos\frac{2\pi}{\lambda}z\cos\frac{2\pi}{\lambda}Vt$$

and a hydrophone to

$$A(f)\cos\frac{2\pi}{\lambda}(z-Vt)-A(f)\cos\frac{2\pi}{\lambda}(z+Vt)=-2A(f)\sin\frac{2\pi}{\lambda}z\sin\frac{2\pi}{\lambda}Vt$$

This shows that for pressure variations there are nodes at $n/2$ and anti-nodes at $(2n+1)/2$ wavelengths below the surface, with $n = 0, 1, 2, \ldots$. Hence, when the predominant signal frequency is 50 Hz and the water velocity 1500 m/s, an offshore streamer should be positioned at 7·5 m below the water surface. Similarly, a geophone should either be placed at the ground surface or 1/2 wavelength deeper, which is 6 m for a frequency of 50 Hz and a weathering velocity of 600 m/s. Fig. 5.4 shows relevant amplitude response curves based on $A_p(f)=\sin\pi f\tau$ and $A_v(f)=\cos\pi f\tau$, with $\tau = 2z/V$.

5.2 CONVOLUTIONS

The reflection coefficient log in Fig. 5.2 will be converted into a reflectivity function $r(t)$ by multiplication with a sequence of properly spaced delta functions. When τ_n is the two-way vertical time to an interface with reflection coefficient r_n, we have

$$r(t) = r_1\delta(t-\tau_1) + r_2\delta(t-\tau_2) + \ldots$$

$$\text{or} \qquad r(t) = \tfrac{1}{2}\sum [\Delta \ln(\rho V)]_n \delta(t-\tau_n)$$

which is the unit impulse reflection response of the earth. For a source signature $w(t)$, the earth response $x(t)$ is the convolution of $w(t)$ and $r(t)$, given by

$$x(t) = r(t) * w(t) = \int_{-\infty}^{+\infty} r(u)w(t-u)\,du$$

Fig. 5.5 is a geometrical illustration of the convolution principle.

The convolution of two time functions corresponds to a multiplication of their frequency spectra. This implies that the outcome of

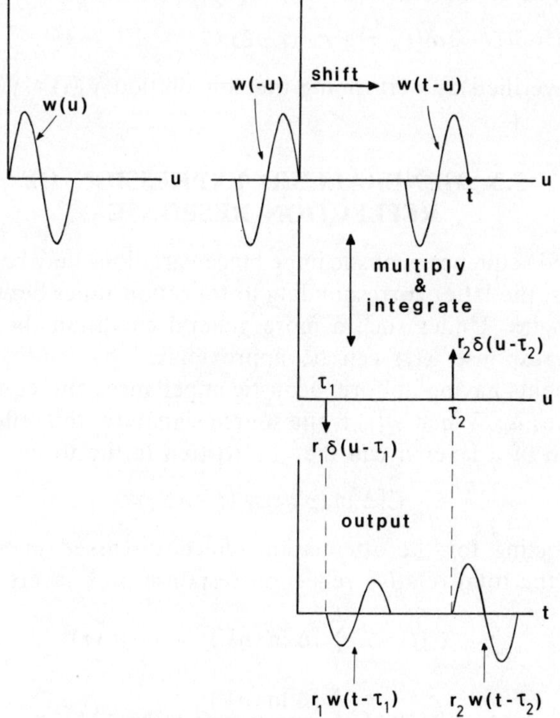

FIG. 5.5. Geometrical illustration of convolution principle.

successive convolutions is independent of the order in which they are performed. The term deconvolution refers to a convolution process which eliminates the effect of a previous convolution. An example of a simple deconvolution filter is the one which cancels the effect of interference between primary reflections and first and higher order peg-leg multiples generated by the water layer during marine surveys (Fig. 3.32). At the shot point and a recording station this reverberating action can be expressed by the unit impulse response

$$h(t) = \delta(t - \tfrac{1}{2}\tau) - r\delta(t - \tfrac{3}{2}\tau) + r^2\delta(t - \tfrac{5}{2}\tau) - r^3\delta(t - \tfrac{7}{2}\tau) + \ldots$$

where r is the reflection coefficient of the sea bottom and τ the two-way water time. Considering the water layer as a set of two filters in tandem, its unit impulse response is the convolution $y(t) = h(t) * h(t)$, or

$$y(t) = \delta(t - \tau) - 2r\delta(t - 2\tau) + 3r^2\delta(t - 3\tau) - 4r^3\delta(t - 4\tau) + \ldots$$

The response of the corresponding deconvolution filter is

$$y_d(t) = \delta(t) + 2r\delta(t - \tau) + r^2\delta(t - 2\tau)$$

as can be verified by performing the convolution $y_d(t) * y(t)$.

5.3 GENERALIZED EXPRESSION OF REFLECTION RESPONSE

In a layered sequence acoustic impedance variations may be discrete or continuous, the latter corresponding to transition zones between different lithologies. Under such a more general condition the subsurface reflection response $x(t)$ can be approximated by considering small layer elements having uniform acoustic impedances and equal two-way transit times $\Delta\tau$. When $w(t)$ is the source signature, the reflection from the bottom of a layer n can then be written in the form

$$C[\Delta \ln(\rho V)]_n w(t - n\Delta\tau)$$

after correcting for the attenuation effects discussed in Section 5.1. Thus, for the total relative reflection response of k layers we have

$$x(t) = C\sum_1^k [\Delta \ln(\rho V)]_n w(t - n\Delta\tau)$$

or

$$x(t) = C\sum_1^k \frac{[\Delta \ln(\rho V)]_n}{\Delta\tau} w(t - n\Delta\tau)\Delta\tau$$

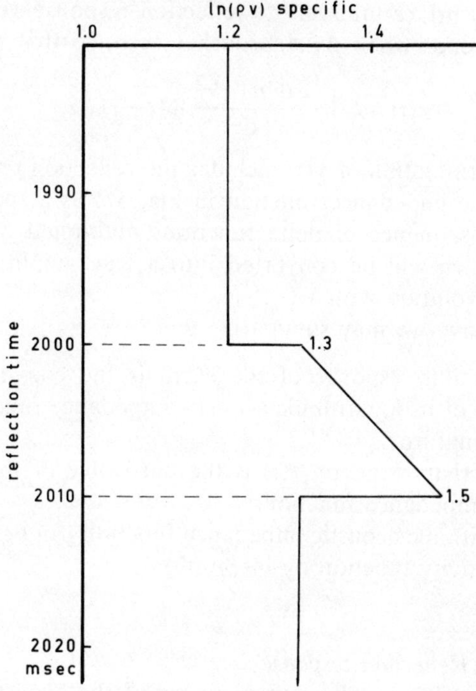

FIG. 5.6. Example 5.2.

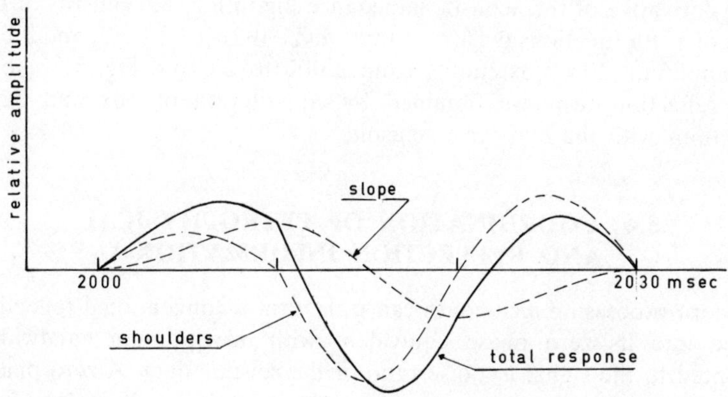

FIG. 5.7. Example 5.2 solution.

For a given subsurface interval the reflection response $x(t)$ is the limit to which $x(t)$ tends when $\Delta\tau$ approaches zero, so that

$$x(t) = C\int_{-\infty}^{+\infty} \frac{\mathrm{d}\ln(\rho V)}{\mathrm{d}\tau}\, w(t-\tau)\,\mathrm{d}\tau$$

This general formulation of $x(t)$ includes the reflection response of the blocked acoustic impedance function in Fig. 5.2 as a special case. Its derivative is a sequence of delta functions multiplied with reflection coefficients, which will be converted into a true amplitude reflection trace after convolution with $w(t)$.

From the above we may summarize that

1. The reflection response of the earth is the convolution of the derivative of its logarithmic acoustic impedance function with the source signature.
2. The reflectivity function $r(t)$ is the derivative of the logarithmic acoustic impedance function.
3. The logarithmic acoustic impedance function can be derived from the reflectivity function by integration.

Example 5.2 *Reflection response*
Determine the relative reflection response of the acoustic impedance profile in Fig. 5.6 to a one-cycle sinusoid of 20 ms duration.

Example 5.2 *Solution*
The derivative of the acoustic impedance log in Fig. 5.6 consists of the pair of delta functions $0{\cdot}1\delta(t-2{\cdot}000)$ and $-0{\cdot}2\delta(t-2{\cdot}010)$, and a box of amplitude $20\,\mathrm{s}^{-1}$ extending from 2·000 to 2·010 s. Fig. 5.7 shows the reflection response obtained by convolution of box and delta functions with the one-cycle sinusoid.

5.4 COORDINATION OF PETROPHYSICAL AND REFLECTION INFORMATION

Modern processing techniques can transform a conventional reflection trace into its zero phase equivalent with an optimum bandwidth, adapted to the signal to noise ratio of the seismic data. A zero phase, true amplitude reflection trace $x_0(t)$ is the convolution of a zero phase wavelet $w_0(t)$ with the reflectivity function of the earth. The wavelet

$w_0(t)$ can be considered as the band limited version of the delta pulse. The integration of $x_0(t)$ will therefore produce the band limited version of the actual logarithmic acoustic impedance variations.

From Section 5.3 we may now distinguish between two alternative ways of comparing seismic information and well data.

1. A conventionally processed reflection trace is matched against a synthetic reflection trace obtained by convolution of a minimum phase wavelet with the differentiated logarithmic acoustic impedance log from the well, after its conversion into two-way vertical time. We have then for the synthetic trace $x(t) = w(t) * Z(t) * D(t)$, where $Z = \ln(\rho V)$ and $D(t)$ is the impulse response of the differentiating filter.
2. A zero phase reflection trace is integrated and compared with the band limited impedance trace from the well. This corresponds to matching $Z(t) * w_0(t)$ against $x_0(t) * I(t)$, $I(t)$ being the integration operator.

Synthetic reflection traces may verify the effectiveness of a multiple elimination procedure, or aid in the identification of reflection events on seismic profiles tied to well locations.

Point 2 above constitutes the basis of lithologic interpretation of seismic reflection data. The outcome of the integration step is illustrated in Fig. 5.8. Trace (a) represents the $\ln(\rho V)$ log derived from petrophysical information. Trace (b) is the corresponding zero phase reflection trace obtained by convolution of a zero phase wavelet with the differentiated $\ln(\rho V)$ time trace. The integration of trace (b) is similar to the summation procedure in Fig. 5.2. The integrated trace (c) is the equivalent of the band limited $\ln(\rho V)$ log and is referred to as a synthetic acoustic impedance log or 'syn log'.

Syn logs are central to an evaluation of reflection amplitude variations in terms of corresponding changes of porosity and porefill of reservoir rocks. Their quantitative interpretation constitutes an important research subject of what is referred to as production seismology. Newly developed assessment procedures based on probabilistic approaches are still of a confidential nature, however, and only partly released for publication (Crans and Berkhout, 1980). From a general point of view, conditions for the effective application of these stratigraphic interpretation methods can be summarized as follows:

1. Discrepancies between well log readings and virgin formation

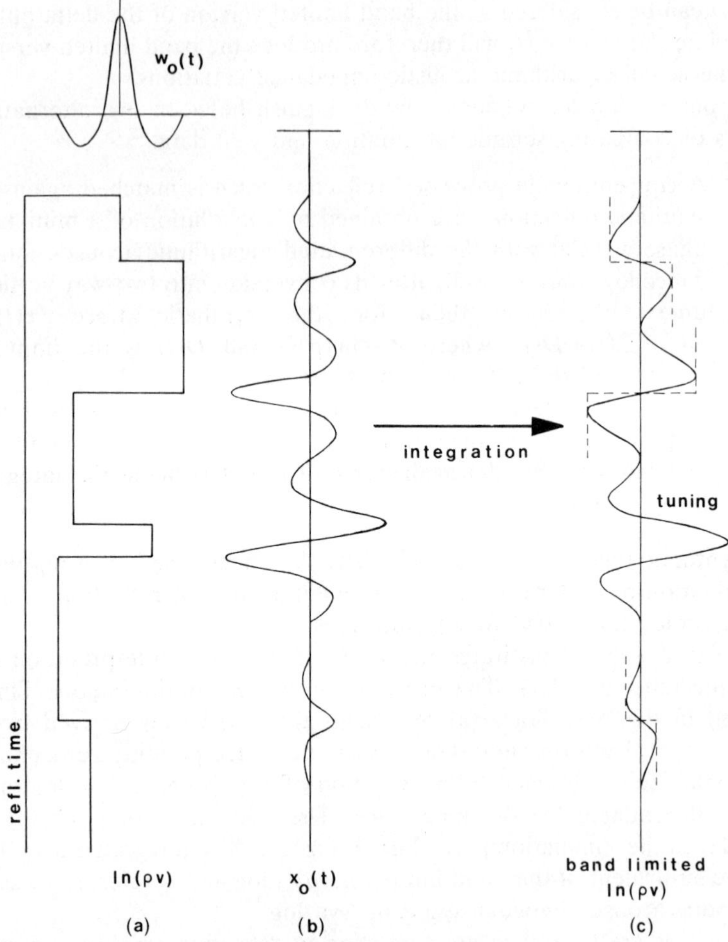

FIG. 5.8. Generation of band-limited ln (ρV) log by integration of zero phase reflection trace.

properties should be eliminated by judicious editing of well log information.

2. Petrophysical logs recorded in deviated wells should be normalized by correcting along-hole logging distance to true vertical depth.
3. Seismic and well data should be compared along the same vertical trajectory at the well site. The seismic information has therefore

to be corrected for migration shifts. This may be a major difficulty in the more complex regions where due to non-uniform strike direction of layer interfaces a two-dimensional migration is not warranted.

5.5 THE GASSMANN EQUATION

The Gassmann equation gives the compressional wave velocity of a porous fluid filled rock (V_{ff}) in terms of its porosity, grain and fluid densities, and various elastic parameters. It plays an important role in the evaluation of reflection amplitude anomalies associated with hydrocarbon accumulations and can be formulated in a compact form as

$$V_{\mathrm{ff}}^2 = \frac{K_s}{\rho}\left[\gamma\beta + \frac{(1-\beta)^2}{1-\beta+B}\right] \qquad \mathrm{cm^2/s^2}$$

where K is the compressibility modulus (bulk modulus) in dyne/cm^2 (1 dyne/cm^2 is $1{\cdot}45\times10^{-5}$ psi), ρ is the bulk density in g/cm^3, given by $\rho = \rho_f + (1-\phi)\rho_s$, ϕ is the fractional porosity, $\gamma = 3(1-\sigma_m)/(1+\sigma_m)$, σ is Poisson's ratio, $B = (K_s/K_f - 1)$ and $\beta = K_m/K_s$. The subscripts f, s and m refer to fluid (oil, gas, brine), solid (the mineral grains of the rock frame) and rock matrix (the empty rock frame) respectively.

It is of interest to note that for zero porosity $\beta = 1$, reducing the Gassmann equation to

$$V^2 = \frac{K_s}{\rho}\gamma$$

from which, upon substitution of $K_s = [E/3(1-2\sigma)]$,

$$V = \left(\frac{M}{\rho}\right)^{\frac{1}{2}}$$

the compressional wave velocity of a solid rock, with M defined as before.

For a mixture of mineral or fluid components appropriate average values for density and bulk modulus are given by

$$\rho_{\mathrm{mix}} = \sum c_i \rho_i$$

$$1/K_{\mathrm{mix}} = \sum \frac{c_i}{K_i} \qquad \text{(Woods' formula)}$$

where c_i is the volume fraction of component i.

The matrix properties K_m and σ_m cannot be determined by direct measurement. Theoretical values for σ_m range from 0 to 0·5. During the evaluation of the Gassmann equation σ_m is often estimated to be of the order of 0·20 or 0·25. Then, when the compressional wave velocity of a specific water saturated rock is measured, its frame bulk modulus K_m can be computed by solving the Gassmann equation for β, giving

$$\beta = 1 - A \pm [(A+B)^2 - B^2/(1-\gamma)]^{\frac{1}{2}}$$

with

$$A = [(\rho V_{\rm ff}^2/K_s) + \gamma(B-1)]/2(1-\gamma)$$

Since the value for K_m is independent of the type of porefill it may subsequently be used to derive the rock's compressional wave velocity for any type of hydrocarbon–brine saturation.

When the shear wave velocity of a rock is known, its shear modulus μ follows from $\mu = \rho V_{\rm sh}^2$. The quantities σ_m and K_m are related to each other by

$$\sigma_m = \frac{3K_m - 2\mu_m}{6K_m + 2\mu_m}$$

whereby it may be assumed that $\mu = \mu_m$. This enables the elimination of σ_m from the Gassmann equation, the rock frame bulk modulus K_m remaining the only unknown parameter.

Example 5.3 *Gassmann equation*

For a water saturated carbonate reservoir it has been found that

Porosity (ϕ)	$= 20\%$
Compressional wave velocity ($V_{\rm wf}$)	$= 4250$ m/s
Grain density (ρ_s)	$= 2{\cdot}78$ g/cm^3
Water density (ρ_w)	$= 1{\cdot}09$ g/cm^3
Water compressibility modulus (K_w)	$= 3{\cdot}05 \times 10^{10}$ dyne/cm^2
Mineral compressibility modulus (K_s)	$= 74{\cdot}5 \times 10^{10}$ dyne/cm^2

Determine

1. The matrix compressibility modulus K_m assuming that $\sigma_m = 0{\cdot}25$.
2. The compressional wave velocity of the gas-bearing carbonate for a water saturation of 25% (S_w), when the gas compressibility modulus (K_g) is $0{\cdot}0428 \times 10^{10}$ dyne/cm^2 and the gas density (ρ_g) 0·157 g/cm^3.
3. The reflection coefficient at the gas–water table.

Example 5.3 *Solution*

1. For the given porosity and grain and water densities the bulk density is 2·44 g/cm^3. We further have for $\sigma_m = 0{\cdot}25$ that $\gamma = 1{\cdot}8$. It then obtains that $\beta = 0{\cdot}2746$ and $K_m = 20{\cdot}46$ dyne/cm^2.
2. Fluid density:

   $$\rho_f = S_w\rho_w + (1 - S_w)\rho_g = 0{\cdot}390 \text{ g/cm}^3$$

   Bulk density:

   $$\rho_{gf} = \phi\rho_f + (1 - \phi)\rho_s = 2{\cdot}302 \text{ g/cm}^3$$

   Bulk modulus of fluid: $K_f = 0{\cdot}0568$ dyne/cm^2, from

   $$1/K_f = S_w/K_w + (1 - S_w)/K_g$$

   Upon substitution of the above quantities in the Gassmann equation we find $V_{gf} = 4000$ m/s.
3. The acoustic impedances of the water- and gas-bearing carbonate are 10 370 and 9208 respectively. The reflection coefficient at the gas–water contact is therefore 0·059.

5.6 SEISMIC EXPRESSION OF HYDROCARBONS

Charging a water-bearing layer with hydrocarbons will increase its absorptive capacity and lower its bulk density and acoustic velocity. The resultant decrease of acoustic impedance depends on depth and on hydrocarbon saturation and can for a given set of physical hydrocarbon and rock properties be evaluated with the aid of the Gassmann equation. Fig. 5.9(a) is a typical example of such acoustic impedance variations for a sand reservoir, from which we may conclude that

1. The influence of hydrocarbon porefill on acoustic impedance diminishes with depth.
2. The effect of gas is consistently larger than that of oil.
3. A low gas saturation causes a relatively large decrease of acoustic impedance.

The change of acoustic rock properties due to the presence of hydrocarbons may be distinguishable on reflection profiles by a diversity of seismic phenomena. These are referred to as hydrocarbon indicators and comprise:

1. *Bright spots and dim spots*, corresponding respectively to a local

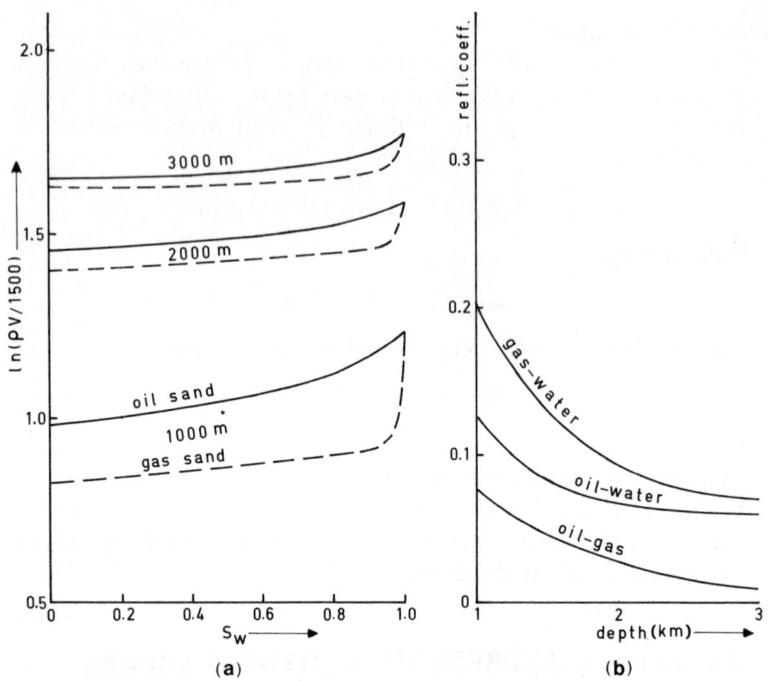

Fig. 5.9. Acoustic impedance of hydrocarbon-bearing sand as a function of depth and water saturation (a). Reflection coefficients at fluid contacts (b).

increase and decrease of reflection amplitudes. The fractional change of reflectivity due to hydrocarbon porefill will decrease with an increase of the acoustic impedance of the sealing formation. Since there is generally only a modest acoustic impedance contrast along sand–shale interfaces, amplitude anomalies associated with hydrocarbon sands encased in shales are always more pronounced than those for sands covered by high impedance seals like salt or anhydrite. For that reason amplitude anomalies in the form of bright spots are exclusively observed in sand–shale sequences and can, from Fig. 5.9(a), be expected to be particularly strong for shallow, low saturation gas sands. Examples of bright spots associated with gas accumulations in Tertiary sands are shown on reflection profiles T, U and V.

A dim spot may be generated by a hydrocarbon sand when the acoustic impedance of its water-bearing part is appreciably larger than that of the sealing shale. A similar dimming of reflection amplitude occurs where hydrocarbons are trapped in high impedance carbonates capped by shales. Reflection profile X shows a good example of such a low amplitude zone along the crest of a gas-bearing limestone reef.

Amplitude variations not related to hydrocarbon accumulations may be caused by lateral changes of lithology (limestone stringers) or by alternating constructive and destructive interferences of reflections from the top and bottom of a thin layer which varies in thickness. Judgement as to the significance of such types of amplitude anomalies in terms of the presence of hydrocarbons must take account of their positions relative to the structural setting in the area of investigation. Another factor is the effect of reflector curvature on reflection amplitude, reflections from culminations and synclinal depressions being registered with diminished and increased intensities respectively. This defocusing and focusing of seismic energy may be compensated by means of an appropriate migration procedure.

2. *Flat spots*, horizontal or near-horizontal alignments of reflection events produced by gas–water, gas–oil and oil–water contacts. These are unconformable with contiguous reflections from the trap boundaries. The perceptibility of flat spots will therefore be enhanced by vertical scale exaggeration. A related phenomenon is an occasional brightening of a flat spot near the edge of a reservoir by its constructive interference with the reflection from the top of the reservoir.

   Fig. 5.9(b) shows theoretical values of reflection coefficients associated with flat spots in hydrocarbon sands. These are of the same order of magnitude as reflection coefficients in water-bearing sand–shale sequences. It is usually the gas–water contacts which produce recognizable flat spots, however. This may be due to the more transitional nature of oil–gas and oil–water contacts. On the other hand, the acoustic impedance contrast along a hydrocarbon–water contact will increase when cementation has taken place in the water-bearing zone of the reservoir. In that case a flat spot may even persist after the reservoir's depletion. Reflection profiles S, R and Q show examples of flat spots related to gas–water contacts.
3. *Inversion of polarity* of a reflection from the top of a reservoir at the edge of a hydrocarbon–water contact. Polarity inversions

occur when the acoustic impedances of the water- and hydrocarbon-bearing parts of the reservoir are respectively larger and smaller than the acoustic impedance of the sealing formation. This implies that polarity changes constitute supporting but not necessary evidence of hydrocarbon accumulations. Examples of polarity inversions of reflections from gas sands are shown on reflection profiles T and V.

4. *Velocity anomalies*, caused by a decrease of the velocity of the hydrocarbon-filled portion of a reservoir relative to the water saturated condition. The low velocity zone may be manifested by a 'sag' or 'pulldown' of reflections from horizons underneath it. Velocity anomalies may be confused with reflector relief related to structure. Judgement in that respect should be based on the consideration that the attitude of a conjectured velocity pulldown must be plausible from a quantitative point of view and consistent in the interval below the low velocity zone. Reflection profile U shows a good example of a time sag underneath a shallow gas accumulation. Doubtful evidence of such a velocity anomaly can be observed below the zone of bright spots on reflection profile T.
5. *Gas chimneys*, columnar regions in the subsurface charged with low concentration gas which has emanated from hydrocarbon reservoirs or other, unidentified sources. Their vertical extent may range from some hundreds of metres to several kilometres. On reflection profiles they correspond to conspicuous shadow zones with a low signal to noise ratio. Faults probably act as migration avenues whereby part of the gas may reach the earth's surface, forming gas seeps. Gas-diffused shales in a gas chimney are characterized by a low velocity and give rise to a strong absorption of seismic energy. Deterioration of reflection response from horizons in and below a gas chimney is believed to be due to the combined effect of stacking difficulties and absorption. Eye-catching evidence of gas chimneys in Tertiary sand–shale sections is shown on reflection profiles U and V. Profile W is an example of a gas chimney which can be associated with a gas-bearing limestone reef. Part of the gas that has leaked through the shale seal has been captured by silt stones encased in the shales, creating a chaotic system of local bright spots. The gas-invaded shale section has caused the partial fade-out of reflection information from the top of the reef and deeper horizons.

Example 5.4 *Amplitude anomaly*
The logarithmic specific acoustic impedances of the gas and brine saturated portions of a sand layer are 0·82 and 1·23 respectively, whereas that of the sealing shale is 1·08. Determine the amplitude ratio of reflections from the gas-filled and the brine-filled parts of the sand layer.

Example 5.4 *Solution*
Reflection coefficient at the shale–brine sand contact: 0·075.
Reflection coefficient at the shale–gas sand contact: −0·130.
Amplitude ratio |1·73|; phase inversion at the edge of the water table.

REFERENCES

Crans, W. and Berkhout, A. J. (1980). Assessment of seismic amplitude anomalies, *Oil & Gas Journal* (Nov. 17).

BIBLIOGRAPHY

Tooley, R. D., Spencer, T. W. and Sagosi, H. F. (1965). Reflection and transmission of plane compressional waves, *Geophysics*, **30,** 552–570.

Muskat, M. and Meres, M.-W. (1940). Reflection and transmission coefficients for plane waves in elastic media, *Geophysics*, **5,** 115–155.

Clewell, D. H. and Simon, R. F. (1950). Seismic wave propagation, *Geophysics*, **15**(1).

Gassmann, F. (1951). Elastic waves through a packing of spheres, *Geophysics*, **16,** 673–685.

Peterson, R. A. (1955). Synthesis of seismograms from well data, *Geophysics*, **20,** 516–538.

Dunlap, H. F., Bradley, J. S. and Moore, T. F. (1960). Marine seep detection: a new reconnaissance method, *Geophysics*, **25,** 275–282.

Geertsma, J. (1961). Velocity log interpretation: the effect of rock bulk compressibility, *Petr. Engr. Journal*, **1,** 235.

Sengbush, R. L., Lawrence, P. L. and McDonal, F. J. (1961). Interpretation of synthetic seismograms, *Geophysics*, **26**(2).

King, M. S. (1966). Wave velocities in rocks as a function of changes in overburden pressure and pore fluid saturants, *Geophysics*, **31**(1).

Gardner, G. H. F., Gardner, L. W. and Gregory, A. R. (1968). Formation velocity and density—the diagnostic basis for stratigraphic traps, *Geophysics*, **39,** 770.

Marr, J. D. (1971). Seismic stratigraphic exploration—Parts I, II, III, *Geophysics*, **36**(2, 3, 4).

Watkins, J. S., Walters, L. A. and Godson, R. H. (1972). Dependence of in-situ compressional-wave velocity on porosity in unsaturated rocks, *Geophysics*, **37**(1).

Domenico, S. N. (1974). Effect of water saturation on seismic reflectivity of sand reservoirs encased in shale, *Geophysics*, **39,** 759.

Backus, M. M. and Chan, R. L. (1975). Flat spot exploration, *Geophysical Prospecting*, **23**(3).

Sheriff, R. E. (1975). Factors affecting seismic amplitudes, *Geophysical Prospecting*, **23,** 125–138.

Anstey, N. A. (1971). *Signal Characteristics and Instrument Specifications*, volume 1 of *Seismic Prospecting Instruments*, Gebrüder Borntraeger, Berlin–Stuttgart.

Gregory, A. R. (1976). Fluid saturation effects on dynamic elastic properties of sedimentary rocks, *Geophysics*, **41**(4).

Toksoz, M. N., Cheng, C. H. and Timur, A. (1976). Fluid saturation effects on dynamic elastic properties of sedimentary rocks, *Geophysics*, **41**(4).

Anstey, N. A. (1977). *Seismic Interpretation: the Physical Aspects*, International Human Resources Development Corporation, Boston, USA.

Payton, C. E. (1977). Seismic stratigraphy—applications to hydrocarbon exploration, Memoir 26, The American Association of Petroleum Geologists.

Kjartansson, E. (1978). The effect of Q on bright spots, 48th Annual Meeting SEG, San Francisco.

May, B. T. and Hron, F. (1978). Synthetic seismic sections of typical petroleum traps, *Geophysics*, **43**(6).

Fertig, J. and Müller, G. (1978). Computations of synthetic seismograms for coal seams with the reflectivity method, *Geophysical Prospecting*, **26**(4).

Maureau, G. T. F. R. and van Wijhe, D. H. (1979). The prediction of porosity in the Permian (Zechstein 2) carbonate of eastern Netherlands using seismic data, *Geophysics*, **44**(9).

Becquey, M., Lavergne, M. and Willm, C. (1979). Acoustic impedance logs computed from seismic traces, *Geophysics*, **44**(9).

Nafi Toksoz, M., Johnston, D. H. and Timur, A. T. (1979). Attenuation of seismic waves in dry and saturated rocks, *Geophysics*, **44**(4).

Dutta, N. C. and Ode, H. (1979). Attenuation and dispersion of compressional waves in fluid-filled rocks with partial gas saturation (White model)—Part I: Biot theory, Part II: Results, *Geophysics*, **44**(11).

Dutta, N. C. and Seriff, A. J. (1979). On White's model of attenuation in rocks with partial gas saturation, *Geophysics*, **44**(11).

Kennett, B. L. N. (1979). Theoretical reflection seismograms for elastic media, *Geophysical Prospecting*, **27**(2).

Taner, M. T., Koehler, F. and Sheriff, R. E. (1979). Complex seismic trace analysis, *Geophysics*, **44**(6).

Ziolkowski, A. (1979). Seismic profiling for coal on land In: *Developments in Geophysical Exploration Methods I*, Applied Science, London.

Brown, R. J. S. and Korringa, J. (1979). Elastic properties of unconsolidated porous sand reservoirs, *Geophysics*, **44**(4).

Mason, J. M., Buchanan, D. J. and Booer, A. K. (1980). Channel waves mapping of coal seams in the United Kingdom, *Geophysics*, **45**(7).

Koefoed, O. and de Voogd, N. (1980). The linear properties of thin layers, with an application to synthetic seismograms over coal seams, *Geophysics*, **45**(8).

Savit, C. H. (1980). Geophysics will find the elusive strat trap, *Oil & Gas Journal* (Nov. 17).

Godfrey, R., Muir, F. and Rocca, F. (1980). Modeling seismic impedance with Markov chains, *Geophysics*, **45**(9).

Sheriff, R. E. (1980). *Seismic Stratigraphy*, International Human Resources Development Corporation, Boston.

Anstey, N. A. (1980). *Seismic Exploration for Sandstone Reservoirs*, International Human Resources Development Corporation, Boston.

Crans, W. and Berkhout, A. J. (1980). Assessment of seismic amplitude anomalies, *Oil & Gas Journal* (Nov. 17).

Bilgeri, D. and Carlini, A. (1981). Non-linear estimation of reflection coefficients from seismic data, *Geophysical Prospecting*, **29**(5).

Meissner, R. and Hegazy, M. A. (1981). The ratio of the PP- to the SS-reflection coefficient as a possible future method to estimate oil and gas reservoirs, *Geophysical Prospecting*, **29**(4).

Balch, A. H., Lee, M. W., Miller, J. J. and Ryder, R. T. (1981). Seismic amplitude anomalies associated with thick First Leo sandstone lenses, eastern Powder River basin, Wyoming, *Geophysics*, **46** (November).

Delaplanche, J., Lafet, Y., Sineriz, B. G. and Remon Gil, H. A. (1982). Seismic reflection applied to sedimentology and gas discovery in the Gulf of Cadiz, *Geophysical Prospecting*, **30**(1).

CHAPTER 6

Velocity Measurements in Wells

6.1 INTRODUCTION

The direct determination of time–depth relationships in deep wells is an important aid to structural and stratigraphic interpretation. Velocity information derived in that manner is superior in accuracy and resolution to that estimated from seismically established stacking velocity distributions.

Detailed information on the earth's velocity profile can be obtained by continuous velocity logging. This involves the recording of formation transit times along small depth intervals by means of an acoustic logging tool which is moved down the borehole. Its basic components comprise a low energy acoustic source which emits in quick succession near-monochromatic pulses with a frequency of the order of 20 kHz, and a pair of receivers at a short distance from the source and separated by a 2 ft interval. The formation transit time is the difference of arrival times of the acoustic signal at the two receivers, expressed in microseconds per foot. Transit times are displayed in the form of a wiggle trace as a function of recording depth, referred to as the sonic log, and progressively integrated to give total time along the borehole interval covered by the logging tool. A continuous velocity log (c.v.l.) can be considered as the reciprocal of the transit time log.

Before the advent of continuous logging tools interval velocities were measured by lowering a single geophone or a pair of geophones into the borehole and recording seismic energy from a source at some distance from the well, employing recording intervals of the order of 100 m. Nowadays well geophone surveys, or 'well shoots', are conducted with the sole purpose of providing a reference time at the starting level of the acoustic survey and to verify the integrated transit times. These auxiliary recordings are usually taken at intervals of several hundred metres and are referred to as check shots and tie shots or calibration shots. During marine well surveys a single hydrophone near the airgun source is commonly used to register a reference time signal on well shoot records. Shot holes on land should be drilled sufficiently deep to avoid weathering correction procedures.

6.2 INTERPRETATION OF WELL SHOOT INFORMATION

The evaluation of well shoot results is based on first arrival times measured on well shoot records. These correspond to travel times along Fermat paths connecting a seismic source with a detector in the well. The general problem is to convert the observed times into along-hole times for comparison with the integrated transit times.

In horizontal velocity distributions well shoot computations can be simplified by using vertical or near-vertical source–detector combinations, so that travel paths can be approximated by straight lines. This implies that for vertical boreholes the seismic source should be placed as close to the well site as safety conditions permit, whereas for deviated holes suitable source positions can be selected from the data of a hole deviation survey. With reference to Fig. 6.1, successive steps for the computation of vertical times below the seismic datum plane are then as follows:

1. Determine the length L of the line connecting a source or gun hydrophone with a well geophone from their respective position coordinates (x_s, y_s, z_s) and (x_g, y_g, z_g) with reference to a fixed point on the derrick floor of the well.
2. Determine the direction angle θ of the ray from $\cos\theta = (z_g - z_s)/L$.
3. Convert a well shoot time T into a vertical time t below the level of the source or gun hydrophone, given by $t = T\cos\theta$.
4. Compute the elevation correction t_e from $t_e = (e_{DF} - z_s)/V_e$, where e_{DF} is the derrick floor elevation with reference to the seismic datum plane and V_e the elevation correction velocity; this may be the water velocity or the velocity of the consolidated formation at datum level for land surveys.
5. Determine the vertical time t_c below the seismic datum plane from $t_c = t - t_e$.

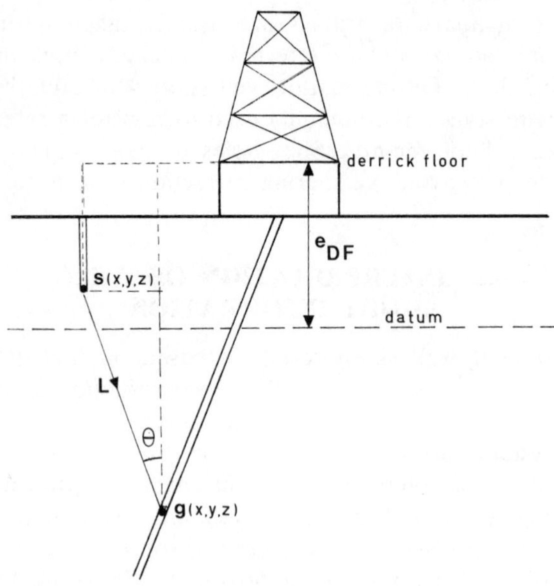

FIG. 6.1. Evaluation of well shoot data in horizontal velocity distribution; system of symbols.

From the vertical times derived above, the travel times along the trajectory of a deviated hole can be calculated. These may then be plotted as a function of the apparent hole depth, together with the integrated transit times.

A more versatile procedure of interpreting well shoot information takes velocity structure and Snell's law into consideration and is based on three-dimensional ray tracing through the sequence of velocity interfaces penetrated by the well. Depending on structural complexity, these can be defined by a set of linear or higher order equations from dip meter evidence and local seismic depth control, using provisional interval velocities provided by the integrated transit time log. The flow chart in Fig. 6.2 shows the main features of a ray tracing program which computes along-hole times from well shoot times. The inner loop on this chart is concerned with finding a Fermat path which connects a geophone in the well with an energy source near the earth's surface. In the intermediate loop this process is repeated in combination with a velocity updating procedure until a match is obtained between computed and observed travel times. The outer program loop

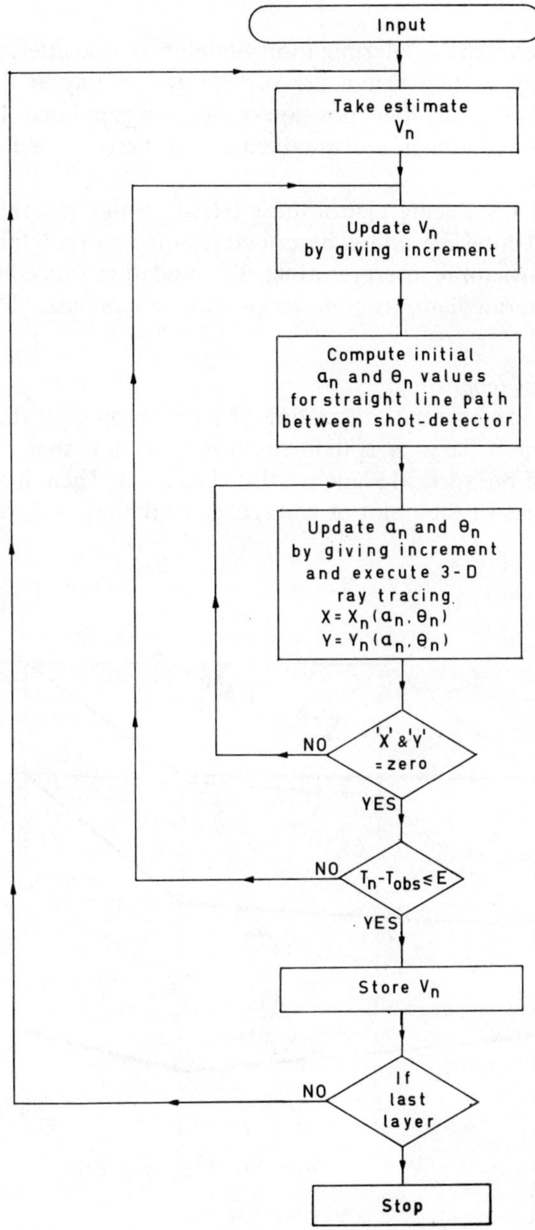

FIG. 6.2. Computation of interval velocities from well shoot times; simplified flow chart.

refers to a sequential determination of interval velocities, starting at layer one immediately below the surface and ending at the velocity layer just above the deepest position of the well geophone. It is hereby assumed that recordings are obtained at the base of each significant velocity layer.

The actual ray tracing algorithm is based on the general principles outlined in Section 3.5 and is described in some more detail in a later section on structural interpretation. The updating procedures in the inner and intermediate program loops can be summarized as follows.

Inner program loop

From Fig. 6.3 the spatial orientation of a ray segment at the level of a well geophone in layer n is defined by its direction angle α_n and the azimuth θ_n of the vertical plane passing through it. Then, if x and y are the coordinates of the point of emergence of the ray with reference to

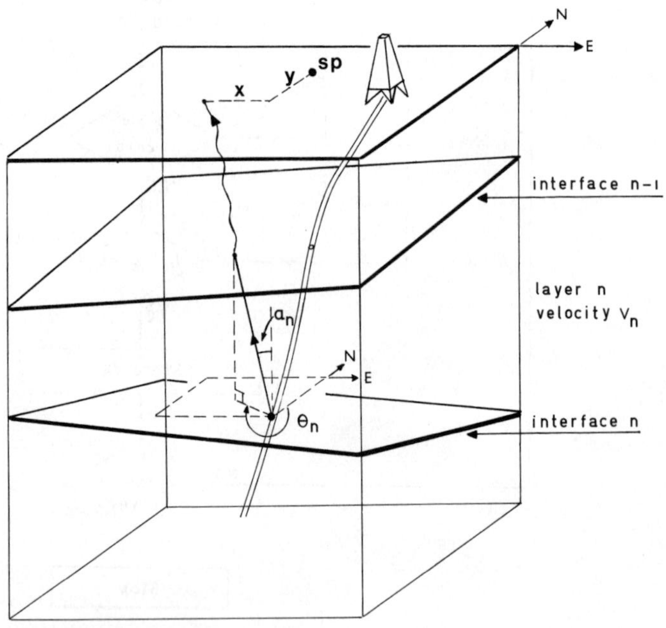

FIG. 6.3. Three-dimensional well shoot geometry; system of symbols.

the relevant seismic source, we have in general

$$x = f_x(\alpha_n, \theta_n)$$

and $$y = f_y(\alpha_n, \theta_n)$$

The problem is to determine the angles α_n and θ_n for which the Fermat path emerges exactly at the position of the seismic source, or for which both x and y become zero. Its solution is obtained by iteration, determining increments $\Delta\alpha_n$ and $\Delta\theta_n$ of α_n and θ_n from the relations

$$x = \frac{\partial x}{\partial \alpha_n}\Delta\alpha_n + \frac{\partial x}{\partial \theta_n}\Delta\theta_n$$

and $$y = \frac{\partial y}{\partial \alpha_n}\Delta\alpha_n + \frac{\partial y}{\partial \theta_n}\Delta\theta_n$$

Initial values of the angles α_n and θ_n correspond to the direction of the straight line connecting the well geophone with the seismic source. Then, during each iteration the partial derivatives in the above equations are derived by the program by varying α_n and θ_n alternately by small amounts. Three or four iterations are usually sufficient to establish the appropriate Fermat path, after which the travel time T_n along it is stored.

Intermediate program loop

In this second loop the velocity V_n of layer n will be updated by increments ΔV_n through a comparison of the observed travel time and the above computed value of T_n. The velocity increments ΔV_n are derived from the relation

$$T_{\text{obs}} - T_n = \frac{\mathrm{d}T_n}{\mathrm{d}V_n}\Delta V_n$$

the derivative $\mathrm{d}T_n/\mathrm{d}V_n$ being determined during each iteration by varying V_n by a small amount. This process is continued until T_n differs less than a preset tolerance from T_{obs}. After this, the value for V_n is stored and the program continues to analyse the velocity layer $n+1$, repeating the entire foregoing procedure.

After completion of the program, interval velocities can be converted into interval times along the borehole trajectory and accumulated times plotted as a function of the apparent hole depth for calibration of the integrated transit times.

Example 6.1 *Well shoot in horizontal velocity distribution*
Determine the effect of neglecting ray curvature on the computation of vertical times to a well geophone at a depth of 1000 m for lateral source–geophone offsets of 100, 200, 300, 400 and 500 m. Zero depth reference at the level of the sources, velocity distribution $V_z = 1600 + 0{\cdot}57z$.

Example 6.1 *Solution*
For the given source positions corresponding ray parameters and travel times can be derived from equations 3.6 and 3.7, the true vertical time to the depth of 1000 m following from equation 3.10. Results are tabulated as follows:

Offset (m)	T (s)	$t_c = T \cos \theta$ (s)	*Error* $\times 10^{-4}$ (s)
0	0·534 6	0·534 6	0
100	0·537 2	0·534 5	1
200	0·545 1	0·534 5	1
300	0·557 9	0·534 4	2
400	0·575 4	0·534 2	4
500	0·597 1	0·534 1	5

Example 6.2 *Effect of velocity structure on well shoot problem*
1. Determine for the model in Fig. 6.4 the well shoot times at detector positions D_1 and D_2 (construct the Fermat path from S to D_2 by a trial and error procedure).
2. Applying a conventional computation scheme for horizontal velocity layering, convert the well shoot times into vertical times and derive the corresponding apparent travel time along the borehole trajectory between D_1 and D_2.
3. Determine the difference between the true and apparent time intervals.

Example 6.2 *Solution*
1. Well shoot time at D_1: 0·611 s.
   Well shoot time at D_2: 0·827 s.
2. Vertical time to D_1: 0·534 s.
   Apparent vertical time to D_2: 0·688 s.
   Apparent time from D_1 to D_2: 0·201 s.
3. True time from D_1 to D_2: 0·238 s.
   Discrepancy with reference to integrated transit times between D_1 and D_2: 0·037 s.

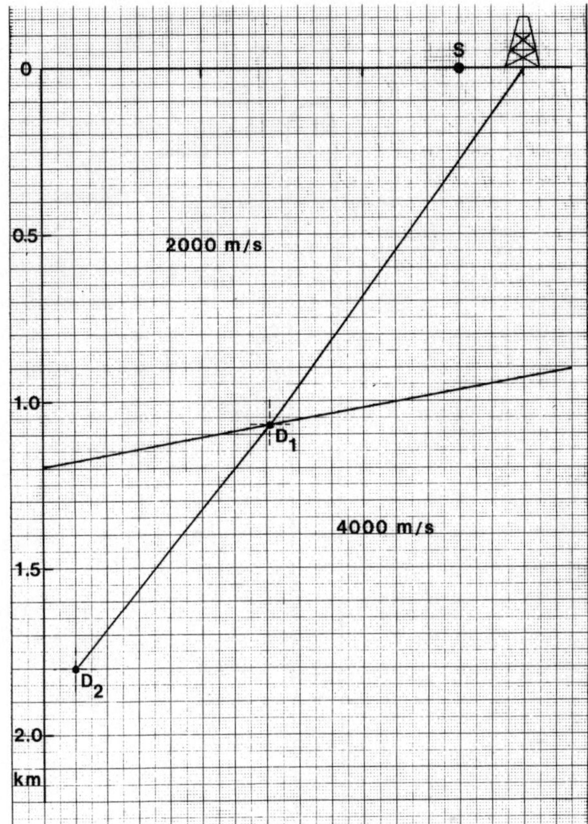

FIG. 6.4. Example 6.2.

6.3 ACOUSTIC LOGGING

Fig. 6.5 shows the principle of acoustic logging. A signal emitted by an energy source at S is refracted in the formation around the borehole and detected by the dual receiver system R_1–R_2. Important dimensions are the separation of the two receivers, called the span, and the spacing, which is the distance from the transmitter to the midpoint of the receiver pair. Refraction ray geometry is similar to that shown in Fig. 3.14, the critical angle of the refracted path corresponding to the ratio of velocities of the drilling mud and the refracting formation. The span of the logging arrangement is normally 2 ft. For the conventional

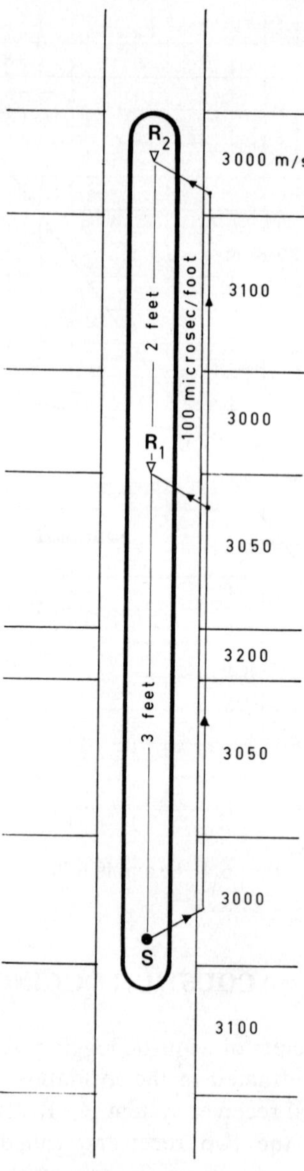

FIG. 6.5. Principle of acoustic logging.

spacing of 4 ft the difference between primary refraction times at receivers R_1 and R_2 is the transit time along a vertical interval of 2 ft in the formation adjacent to the borehole wall. Expressed in microseconds per foot, its reciprocal is the average P-wave velocity along the 2 ft formation interval.

Tools in current use measure transit times in directions up and down the borehole to eliminate errors due to sonde tilt and hole size variations. From Fig. 3.15, this is analogous to up- and downdip shooting during refraction work. The borehole compensated tool (b.h.c.) contains two transmitters and two sets of receivers which cover the same formation interval. Transit time values from the two receiver pairs are averaged automatically during the logging procedure.

On the sonic log the transit time scale normally extends from 40 to 200 μs/ft, which corresponds to a velocity range of 7600 to 1500 m/s. The integrated transit times are given by a sequence of integration pips along the depth axis of the log, each pip indicating an increase of 1 ms of the total travel time along the logged borehole interval. From this information a detailed time versus depth plot can be obtained, referenced to the starting position of the sonic survey. The integrated times are adjusted to times below the seismic datum plane by calibration with the set of reference times from the well geophone survey. An example of a unified T–Z plot for these two types of data sets is shown in Fig. 6.6.

Various conditions cause measured transit times to differ from true transit times in the virgin formation near the borehole. This may impede the comparison of the integrated times with calibration times as well as that of the log-derived acoustic impedances with seismic acoustic impedance amplitudes. Substantial discrepancies between integrated transit times and well shoot results have been observed for instance along shale intervals. These are due to swelling of the shale zone immediately around the borehole. The conventional sonic logging tool measures the relatively low velocity of the altered shale zone and it was found that higher shale velocities were registered by increasing the tool's spacing. This has prompted the development of a logging sonde with a spacing of the order of 10 ft. At this larger recording distance refractions from the virgin shale zone arrive as primary events, in analogy with the three-layer refraction geometry in Fig. 3.17. Other causes of anomalous transit times are

1. Attenuation of the acoustic signal due to geometric spreading in

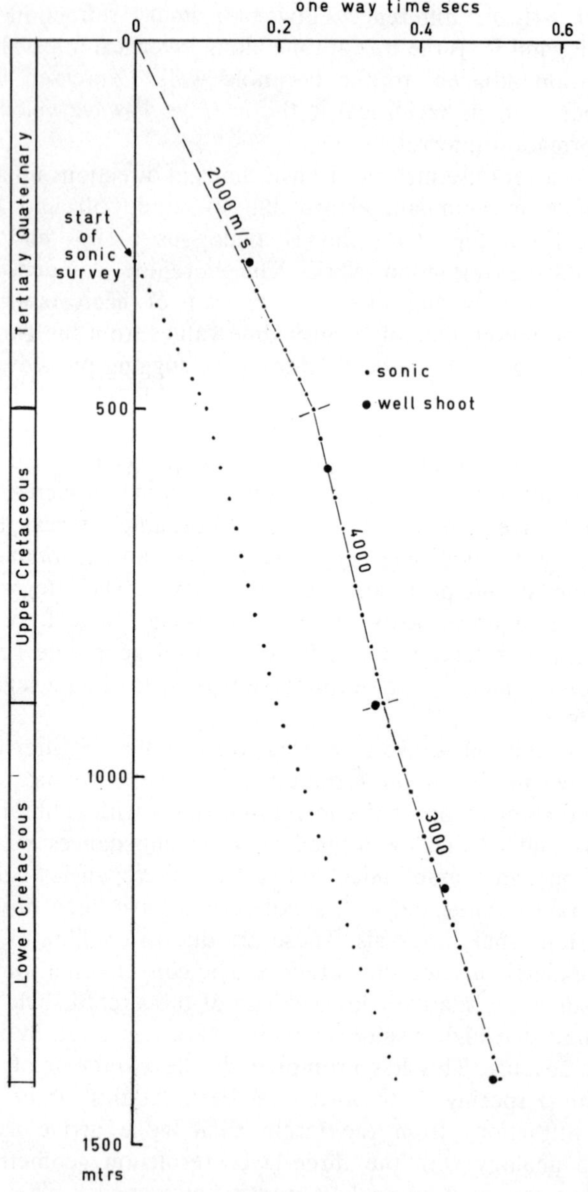

FIG. 6.6. Example of t–z plot based on well shoot and sonic data.

caved parts of the hole, by scatter in fractured formations or by absorption in gas-bearing layers. The onset of the first cycle of the sonic wave train may then be too weak to trigger the far-distant receiver. Instead, a later cycle of higher amplitude is detected, leading to the registration of too large transit time values on the log; this effect is called cycle skipping.

2. Noise interference, which may cause the premature triggering of both receivers. There is a slightly larger chance that this will affect the far-distant receiver. In that case transit time readings will be too low.
3. Invasion of mud filtrate into permeable hydrocarbon bearing strata. This will increase their velocities in a narrow zone around the borehole. Transit time values will therefore be too low. Since velocities decrease in the radial direction such an effect cannot be remedied by the long-spaced sonic tool.

Prior to the generation of acoustic impedance logs, discrepancies mentioned in points 1 and 2 should be eliminated by estimating the true transit time values from interlog relationships, such as density–velocity cross-plots. The editing for invasion effects involves the computation of the true *in situ* velocities of hydrocarbon-bearing layers with the aid of the Gassmann equation.

Another point of interest with regard to seismic problems is the resolution of a continuous velocity log. Fig. 6.5 indicates that an acoustic sonde measures the average formation velocity along an interval of 2 ft. A continuous velocity log can thus be considered as the convolution of the actual velocity–depth relationship with a box of unit amplitude and a width of 2 ft. After conversion into two-way time for a uniform velocity of 3000 m/s, this becomes equivalent to convolution with a box of a duration of 0·4 ms. From Section 2.13 the corresponding low pass filter has a frequency cutoff near 2500 Hz, which amply covers the range of seismic frequencies. The fidelity of acoustic impedance logs appears therefore sufficient for seismic amplitude calibration.

BIBLIOGRAPHY

Slotnick, M. M., Brooks, J. A. and Redding, V. L. (1950). A note on well-shooting data and linear increase of velocity, *Geophysics*, **15**(4).

Levin, F. K. and Lynn, R. D. (1958). Deep-hole geophone studies, *Geophysics*, **23**(4).

Dunoyer de Segonzac, Ph. and Laherrere, J. (1959). Application of the continuous velocity log to anisotropy measurements in Northern Sahara; results and consequences, *Geophysical Prospecting*, **7**(2).

Wyrobek, S. M. (1959). Well velocity determinations in the English Trias, Permian and Carboniferous, *Geophysical Prospecting*, **7**(2).

Kokesh, F. P., Schwartz, R. J., Wall, W. B. and Morris, R. L. (1965). A new approach to sonic logging and other acoustic measurements, *Journal Pet. Tech.*, **17**(3).

Kennett, P. and Ireson, R. L. (1971). Recent developments in well velocity surveys and the use of calibrated acoustic logs, *Geophysical Prospecting*, **19**(3).

Bais, P., La Porte, M., Lavergue, M. and Thomas, G. (1972). Well-to-well seismic measurements, *Geophysics*, **37**(3).

Chander, R. (1977). On tracing seismic rays with specified end points in layers of constant velocity and plane interfaces, *Geophysical Prospecting*, **25**(1).

Kennett, P. (1979). Well geophone surveys and the calibration of acoustic velocity logs. In: *Development in Geophysical Exploration Methods I*, Applied Science, London.

Kitsunezaki, C. (1980). A new method for shear wave logging, *Geophysics*, **45**(10).

CHAPTER 7

Structural Interpretation of Reflection Information

7.1 INTRODUCTION

The structural evaluation of reflection data is concerned with the solution of two basic problems. The first one is the creation of a reflection time image of the subsurface by tracing and mapping an appropriate set of reflection horizons, which is often a matter of geological rather than geophysical insight. The second one is the conversion of the time domain solution into depth. During the early days of reflection prospecting a substantial part of the total interpretational effort was devoted to the correlation of reflection events along sequences of wiggle field records and the plotting of hand-picked reflection times below shot–detector midpoint positions on graph paper, a procedure referred to as vertical plotting or 'point plotting'. This tedious manual work has now been entirely superseded by the digital processing of multi-coverage reflection information, considerably facilitating the present-day time domain interpretation and enhancing its precision and resolution. In many instances dependable depth conversion remains the principal problem, however; the solution may be hampered by complex velocity structure, difficulties in evaluating migration effects, paucity of velocity information and the occasional distortion of stacked signal alignments.

The automatic migration of reflection profiles is only trustworthy as an interpretational aid when the reflection traverse extends in the direction of subsurface dip. Structural interpretation in the depth domain is therefore basically a three-dimensional problem where approximations from vertical depth conversion are unsatisfactory.

Fidelity of stacked information depends on to what extent the normal incidence condition has been fulfilled after moveout corrections. In that respect it is noteworthy that post-stack migration systems are strictly based on a normal incidence reflection geometry. During the time domain interpretation there is not always tangible evidence of deviations of stacked signals from their theoretical normal incidence positions. These discrepancies may manifest themselves, however, as spurious structural relief after depth conversion. In areas of low to moderate complexity such complications are generally of local significance and will not seriously affect the overall accuracy of structural interpretations. Under more adverse conditions of highly disturbed velocity overburdens, however, a meaningful interpretation of stacked information from underlying targets often becomes questionable. A solution in the depth domain may then be obtained by the pre-stack migration of c.m.p. gathers, corrected only for static shifts. This will be further discussed in Chapter 8.

7.2 TIME DOMAIN INTERPRETATION

General proceedings

These are summarized as follows

1. Identification of reflection horizons by tying seismic traces to time converted well log data, by correlation with interpreted reflection information along contiguous or intersecting program grids, or from the interpreter's experience in the general exploration area.
2. Specification of prospective intervals.
3. Investigation of seismic evidence of hydrocarbon occurrence.
4. Analysis and coordination of fault indications along reflection profiles; preparation of compilation maps showing main fault patterns, outlines of structural trends and the positions of special features like subcrops and diapirs.
5. The tracing of a representative set of reflection horizons. These comprise bedding planes and non-angular unconformities distinguished by consistency of reflection character, disconformities along which reflection character may vary, and phantom lines drawn parallel to structural event alignments of limited lateral extent.
6. Digitization of reflection profiles. This is the semi-automatic registration of reflection times for the interpreted events. A reading device called a cursor is moved along the colour-coded

horizons after fixing the reflection section to a digitizing table. The cursor is triggered by the table operator to register table coordinates at prescribed lateral intervals. These are printed on magnetic tape or cards. Table coordinates are referenced to the horizontal and vertical scales of the reflection section and later converted into shot point or trace coordinates and reflection times. This information is stored to serve as input to computer programs for listing it in a convenient format, for plotting posted reflection time maps and for performing depth conversions.

Limitations of stacked information

With reference to preceding sections the principal deficiencies of stacked reflection profiles in comparison with results from normal incidence modelling are recapitulated below:

1. Lack of shallow reflection information.
2. Attenuation of steep dip events from fault planes and flanks of diapirs where stacking velocities are adapted to sedimentary dip.
3. Attenuation of the steeper parts of diffraction alignments.
4. Distortion or fade-out of reflection information from horizons underneath complex velocity structure.

Fig. 7.1 illustrates the effect mentioned in point 4 on a small scale. Non-hyperbolic moveouts along c.m.p. profiles covering both the up- and downthrown fault blocks cause a gradual shift of stacked signals relative to their normal incidence positions in the lateral interval between the undisturbed c.m.p. profiles. This results in flexure of the deeper horizon underneath the fault, where normal incidence ray tracing would produce a time discontinuity. Such data constitute invalid inputs to post-stack migration systems, while vertical depth conversion will lead to false structural relief, which may be recognized and eliminated during the construction of depth contour maps.

Time ties at line crossings

Since normal incidence times are scalar quantities it is evident that reflection traces belonging to different profiles should match at their intersection. This property is an important interpretational aid in identifying and correlating reflection events along a grid of seismic traverses. When reflection alignments terminate at a short distance from the intersection of traverses, the tie condition should be fulfilled by their extrapolation in the direction of the line crossing. This

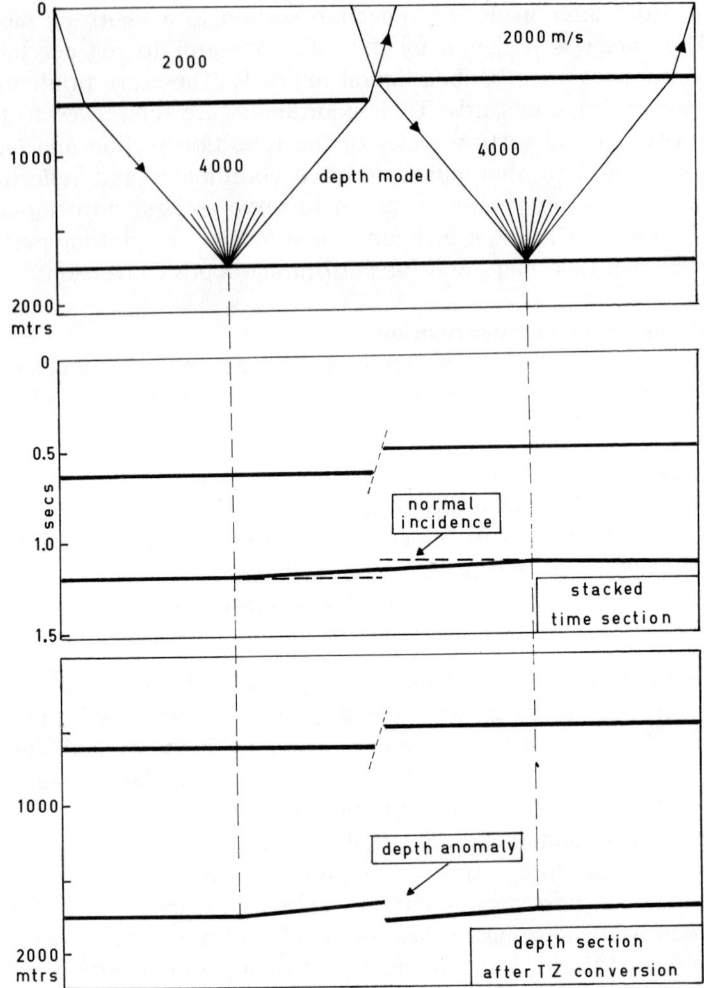

FIG. 7.1. Depth conversion errors through effect of lateral overburden velocity discontinuity on attitude of deeper stacked horizon.

procedure provides an effective check on the correctness of event correlations and is illustrated in Fig. 7.2 for the following cases:

1. Truncation of a reflecting interface by a fault near the intersection of two traverses. Depending upon the direction of throw of the fault, extrapolated time ties should be established for reflection horizons from either up- or downthrown fault blocks.

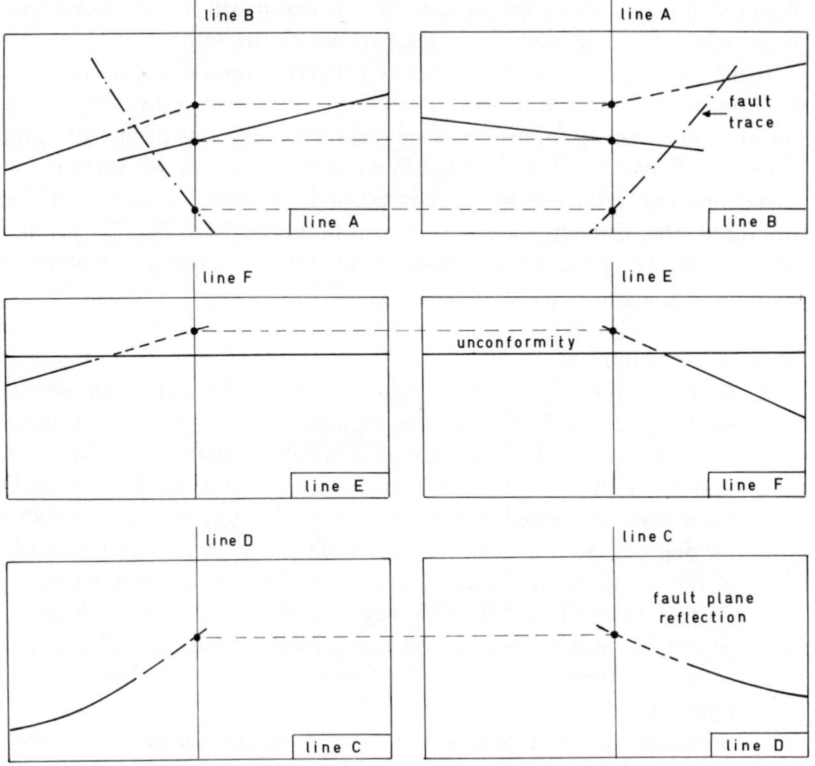

FIG. 7.2. Time ties at line crossings.

2. Truncation of a reflecting interface by an unconformity near the intersection of two traverses.
3. Terminations of fault plane reflections near the intersection of two traverses. The manifestation of these events depends on the acoustic impedance contrast along the fault plane and on stacking response. The latter will generally be larger for profiles parallel to the fault than for dip profiles. Due account of this processing aspect should be taken when correlating steep dip events at line crossings.

There may be a perceptible mistie when on either or both of the intersecting profiles signal alignments deviate from their normal incidence positions. From Fig. 7.1 it can for instance be inferred that the flexured interval along the deeper time horizon will not tie to normal incidence reflections along profiles parallel to the fault.

It is further an important aspect of reflection interpretation that the tie condition at line intersections should also be met by any of the interpreter's constructions showing estimated true positions of fault traces or flanks of diapiric intrusions. With reference to Section 7.4 these are the intersections of corresponding spatial features in the migrated time domain with the vertical plane of the reflection profile. The fulfilment of the tie condition is in this case clearly a matter of maintaining a geometrical consistency.

Illustrative examples

1. Sections B, C and D are sixfold marine profiles across an anticlinal feature in a Tertiary clastic sequence of the Niger Delta basin. Lines B and C, 3·5 km apart, extend in a north–south direction and cover the fault-bounded north flank of the structure. Line D is an east–west profile which crosses the fault at an angle of about 30 degrees. Reflections from the fault plane tie straightforwardly at the intersection of lines C and D, and through extrapolation at that of lines D and B. This suggests that the three fault plane reflections are related to the same fault. From this inference a reflection time contour map of part of the fault plane can be constructed.
2. Sections G, H, I and J are sixfold stacks, offshore Norway, covering a cylindrical salt diapir in a near-horizontal sequence of Tertiary and older rocks. The base of the Tertiary is at about 3·5 s. Fig. 7.3 is a situation map showing reflection time contours of the cap rock of the diapir. Along some of the profiles the fade-out of reflections from its steep flank may be due to the combined effect of too low a stacking velocity and the filtering action of detector patterns. Note in that respect the difference in amplitude of cap rock reflections along lines I and J near their intersection. Direct ties of the cap rock reflections can be established at the intersections of line H with lines G and I, whereas extrapolated ties are inferred at the crossings of lines J and I and lines J and G. Of special interest is line J, which manifests the reflection response of the cap rock in a so-called side swipe position. By tying line J to line I it can further be concluded that the steep dip event along line J at about 4 s near its intersection with line I is due to diffraction from the circumference of the salt diapir.

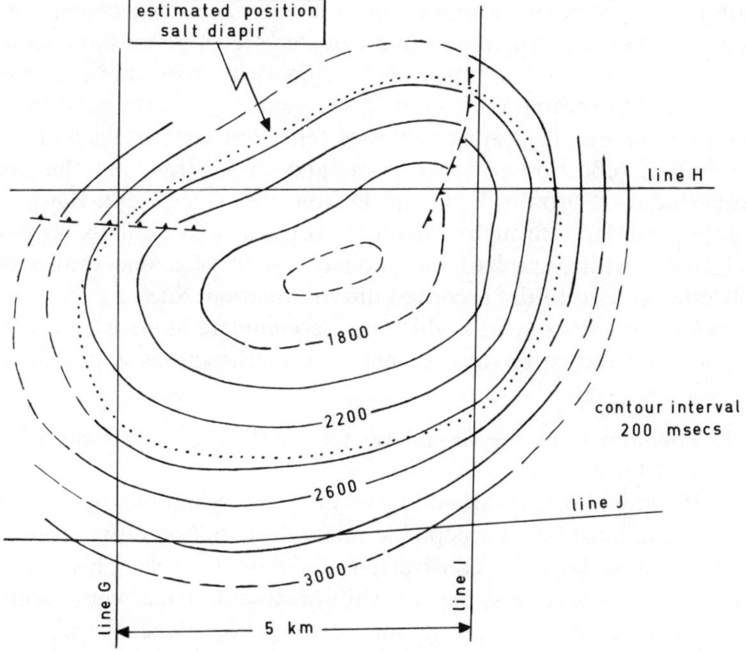

FIG. 7.3. Reflection time contour map of cap rock salt diapir.

3. Sections K, L and M are 24-fold North Sea profiles near the Dutch coast. The feature of interest is a local dolomite deposit, shaped in the form of an anticline and embedded in Zechstein salt. It is trending roughly NW–SE and is crossed in an east–west direction by lines K and L. Line M extends north–south, about 1 km to the east of the southeastern termination of the dolomite layer. The strongly curved reflection alignment along line M is again a typical side swipe. It is the reflection response of the southeastern plunge of the anticline as follows from ties to profiles K and L.

Phantom horizons

In sand–shale sequences lateral facies changes due to changes of the environment of deposition may cause corresponding variations of reflection character, limiting the lateral extent of correlative reflection events. Structural mapping can then be based on the tracing of phantom horizons, which are lines conformable to nearby reflection patterns indicative of structural dip. A phantom can thus be considered as a record of the structural attitude at some level of constant geologic time. This implies that phantom horizons will cross transgressive or regressive lithostratigraphic boundaries, generally not revealed by seismic information. Typical examples of reflection sections showing non-continuous reflection patterns from sand–shale strata are the previously discussed profiles B, C and D from the Niger Delta Basin.

Interpretation through phantom constructions entails personal judgement with regard to the goodness of fit of a line drawn on a reflection profile to the recorded dip information. Such a procedure is therefore subject to errors which may accumulate along a profile grid. A check on the correctness of phantom constructions is provided by the following criteria:

1. Phantom horizons should tie to identical structural markers at well locations.
2. Phantom horizons should tie at all line crossings of a profile grid.
3. There must be an acceptable relationship between the attitude of phantom horizons constructed at different levels. This pertains, for instance, to a systematic shift of structural axes with depth, or to a regular divergence or convergence along the flanks of anticlinal features.

Misties or phantoms at line intersections can be expressed in terms of errors of closure along profile loops. The registration of these discrepancies on loop closure charts will greatly facilitate the determination of an optimum strategy of adjusting initial estimates of phantom slopes. Changes of reflection character across major faults of the type shown on profiles B and C prevent a straightforward correlation of phantoms in up- and downthrown fault blocks. In such cases sets of phantom horizons in the individual fault blocks are independent structural entities. Each phantom is then only distinguished by its position—shallow, intermediate or deep—in the sedimentary section, unless identified at well locations. Phantom constructions across secondary crestal or antithetic faults may be based on local correlation of reflection character, perhaps supported by reference to migrated information. A substantial part of the minor faults remains undetected, however. Therefore, phantom horizons should in general be regarded as smoothed versions of actual structural configurations. This often becomes apparent in hydrocarbon-bearing regions from discrepancies between contour maps of phantom horizons and more detailed structural maps based on well information.

7.3 NORMAL INCIDENCE REFLECTION GEOMETRY

Considering a reflecting interface as a fictive source of seismic energy, upgoing wave fronts can be thought of being surfaces of equal two-way travel time oriented at right angles to its normal incidence ray system. This implies that normal incidence rays terminate perpendicularly to reflection time contours along a corresponding reflection horizon. Fig. 7.4 is an illustration of this relationship, which leads to the important

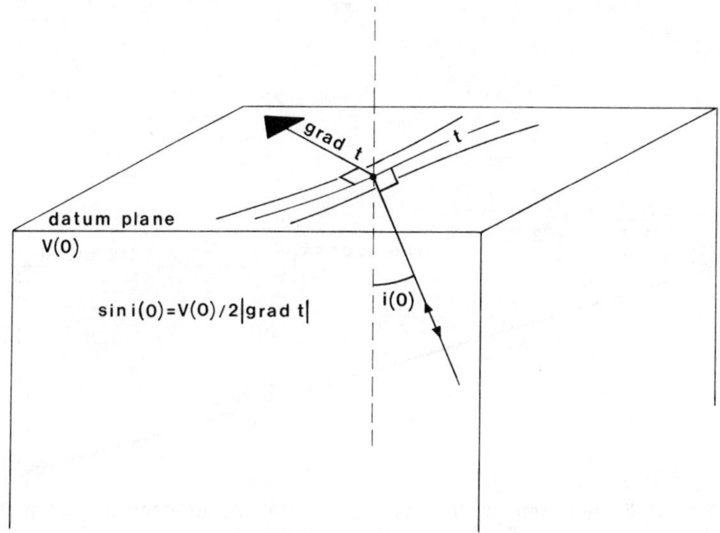

FIG. 7.4. Relationship between emergence angle of normal incidence reflection path and reflection time gradient.

property that the upper segment of a normal incidence ray is contained in the vertical plane through the reflection time gradient vector at its point of emergence. From Section 2.10 the angle of emergence $i(0)$ is given by

$$\sin i(0) = \frac{V(0)}{2} |\text{grad } t|$$

where t denotes two-way travel time.

After expressing the initial orientation of a normal incidence ray in terms of a set of direction cosines it can, from Section 3.5, for a specific reflection time be traced through a given velocity distribution until it ends at a reflection point. This is the basic principle of the migration of reflection time contour maps.

Time dips along a reflection profile are the components of time gradient vectors in the direction of the profile. From Fig. 7.5 the reflection time gradient of a reflection horizon can at any point in the

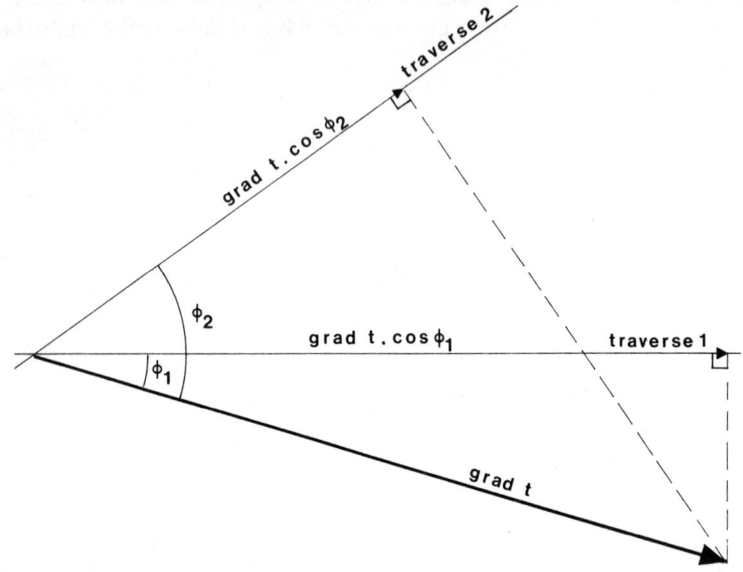

FIG. 7.5. Reflection time gradient derived from its components along a pair of reflection traverses.

seismic datum plane be derived from its components measured along a pair of arbitrarily oriented traverses. Similarly, we have for a normal incidence time field $t(x, y)$

$$\text{grad } t = \frac{\partial t}{\partial x}\mathbf{i} + \frac{\partial t}{\partial y}\mathbf{j}$$

where **i** and **j** are unit vectors in the directions of the X- and Y-axes.

Example 7.1 *Structural dip, emergence angle, time gradient*

The structural dip of a reflecting interface is 30°. Determine for a velocity distribution $V_z = 1500 + 0{\cdot}5z$ m/s:

1. The angle of emergence of a normal incidence ray from a reflection point at a depth of 1000 m.
2. The reflection time gradient at the point of emergence of the normal incidence ray.

Example 7.1 *Solution*

1. From Snell's law: $\sin i(0) = (1500/2000) \sin 30° = 0{\cdot}375$, so that $i(0) = 22°$.
2. $|\text{grad } t| = [2/V(0)] \sin i(0) = 5 \times 10^{-4}$ s/m.

Example 7.2 *Scales, time gradient, emergence angle I*

A reflection profile has a horizontal scale of 1:25 000 and a vertical scale of 80 mm for 1 s. The slope of a reflection segment as measured with a protractor is 45°. Determine

1. The time dip in s/km.
2. The angle of emergence of the normal incidence ray, assuming that the profile extends in the direction of subsurface dip.

Velocity distribution: $V_z = 2000 + 0{\cdot}7z$ m/s

Example 7.2 *Solution*

1. The time dip is 0·5 s/km.
2. For the near-surface velocity of 2000 m/s the emergence angle is 30°. The velocity gradient 0·7 is redundant information.

Example 7.3 *Scales, time gradient, emergence angle II*

Fig. 7.6 shows three reflection horizons on a normal incidence profile extending in the direction of subsurface dip. Determine the angle of emergence of the normal incidence ray from the second horizon.

Example 7.3 *Solution*

The time dip along horizon 2 is $6{\cdot}8\ 10^{-5}$ s/m. The near-surface velocity is 2000 m/s, so that $\sin i(0) = 0{\cdot}068$ and $i(0) = 3{\cdot}9°$. Through its presentation in the form of a t–x profile it is often not recognized that the present exercise is virtually a duplication of the previous one. The 3000 m/s velocity and the time dip along horizon 1 are irrelevant to the solution of the problem.

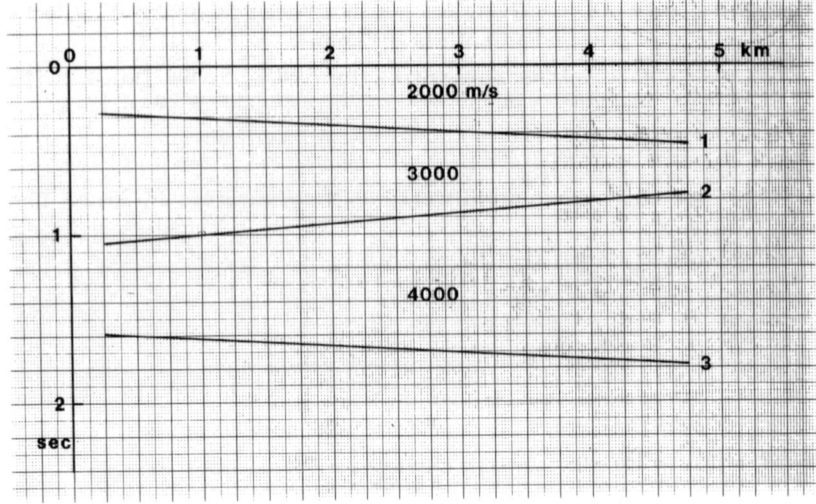

FIG. 7.6. Example 7.3.

7.4 GLOSSARY OF MIGRATION TERMS

In reflection work migration is the determination of the true spatial positions of reflecting interfaces from their manifestation in the reflection time domain. Inverse migration is the transformation of a geological model into its reflection time image. Basic entities in migration geometry can be summarized as follows

1. *Migrated depth*, the depth of a point on a reflecting interface below the seismic datum plane.
2. *Migrated time*, migrated depth converted into two-way time. When depth intervals measured along the vertical through a sequence of reflectors are $(\Delta z)_1, (\Delta z)_2, \ldots, (\Delta z)_n$ and corresponding interval velocities $V_1, V_2, \ldots, V_n$, the migrated time to the nth reflector is $2 \sum_n (\Delta z)_n / V_n$. The onset times of reflection signals along a migrated time trace are identical to the migrated times of corresponding reflection points.
3. *Unmigrated time*, the two-way time of a reflection signal along a normal incidence reflection trace. For non-zero offset data in horizontal velocity distributions unmigrated time is defined as the

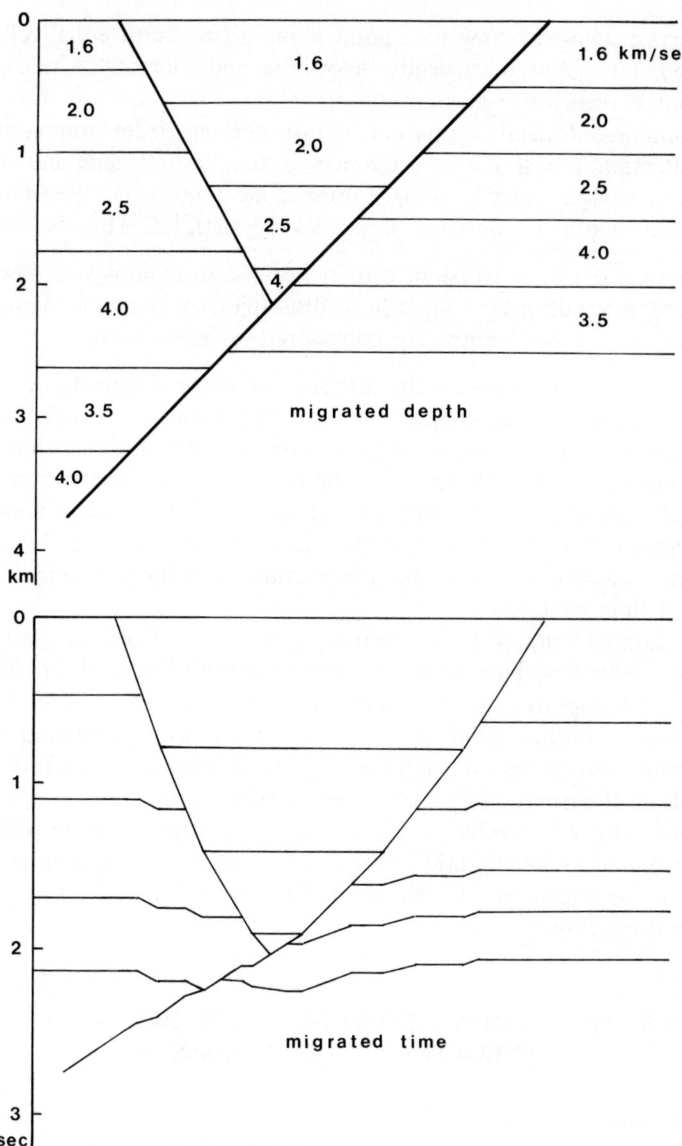

FIG. 7.7. Migrated time derived from migrated depth.

vertical two-way time to a point along a surface of equal reflection time, plotted vertically below the midpoint of the relevant source–receiver pair.

4. *Unmigrated depth,* apparent depth derived from unmigrated reflection times. For a sequence of unmigrated time intervals $(\Delta t)_1, (\Delta t)_2, \ldots, (\Delta t)_n$ along a normal incidence trace the unmigrated depth to horizon n equals $\frac{1}{2}\sum_n (\Delta t)_n V_n$. This is called vertical depth conversion, usually applied in regions with low to moderate subsurface dip where discrepancies between migrated and unmigrated depths are considered as insignificant.

On reflection time profiles the attitudes of inferred fault traces and other non-registered geological elements are usually in conflict with the above definition of migrated time. This is due to the fact that the interpreter's concept with regard to the true shape of these features in the depth domain is commonly copied in the migrated time domain with disregard of the effects of the scale conversion. Fig. 7.7 is a synthetic example showing the incongruity of migrated depth and migrated time configurations.

Migration techniques derive from ray geometry or from wave acoustics. The geometrical approach is concerned with the two- or three-dimensional migration of a limited number of horizons after their interpretation on time sections; this is referred to as horizon migration or contour map migration. Migration based on wave acoustics involves the shift of all signals along a sequence of reflection traces to migrated positions. This will produce migrated depth or migrated time profiles by two-dimensional migration along dip lines, or a spatial migrated image of the subsurface by the three-dimensional migration of areal reflection coverage.

7.5 MIGRATION GEOMETRY IN HORIZONTAL VELOCITY DISTRIBUTIONS

Introduction

Isovelocity planes in continuous velocity distributions are generally unconformable to structural bedding planes. Such a relationship between velocity distribution and geologic structure is typical for clastic sequences where average velocity regimes are shaped under the influence of compaction and conform to the trends of continuous velocity logs. A reflection process on the other hand is controlled by the fine details of density and velocity variations so that reflection time gradients may be commensurate to structural dips measured by correlation of transit time signature. When the velocity distribution is horizontal there will be a close correspondence between the t–z graphs and between the trends of transit time logs from a set of well locations; likewise, refraction t–x curves will show uniform shapes. Compaction of sediments is then only dependent on their present depth of burial, whereas structural relief is mainly caused by differential subsidence during deposition. Since compaction is an irreversible process, tectonic inversion may fold or tilt a velocity distribution that was originally horizontal, inducing velocity structure. When its relief is moderate, migration algorithms may in restricted regions be based on a horizontal velocity distribution model and lateral velocity variations accounted for by appropriate changes of the velocity distribution parameters.

Wavefront charts

In migration geometry a wavefront chart is defined as a set of curves of equal two-way travel time with reference to a coincident location of seismic source and receiver. In a horizontal velocity distribution these curves are symmetrical about the vertical depth axis passing through the origin of the wavefront chart. A rotation of the chart around its axis will therefore produce surfaces of equal two-way travel time.

The above describes a wavefront chart in the depth domain, usually referred to as a *ZX* chart. A change of the depth scale into a two-way time scale will transform a *ZX* chart into a *TX* chart. An example of a *TX* wavefront chart for a linear velocity distribution is shown in Fig. 7.8. From Section 3.4, the wavefront curves on this chart correspond to circles in the *ZX* domain with centres distributed along the depth axis.

The axis of a *TX* wavefront chart is the locus of unmigrated (or 'vertically plotted') reflection times along a normal incidence trace at the origin of the chart. For a normal incidence time t, unmigrated points on *ZX* and *TX* charts are plotted where a wavefront of time t cuts their vertical axes. The corresponding migrated reflector element is tangent to the wavefront of time t in *ZX* and *TX* domains. The point of tangency represents a migrated reflection point, situated at the intersection of wavefront t with the normal incidence ray specified by the time gradient of the unmigrated reflector element.

Since the velocity normally increases with depth, the inclination of

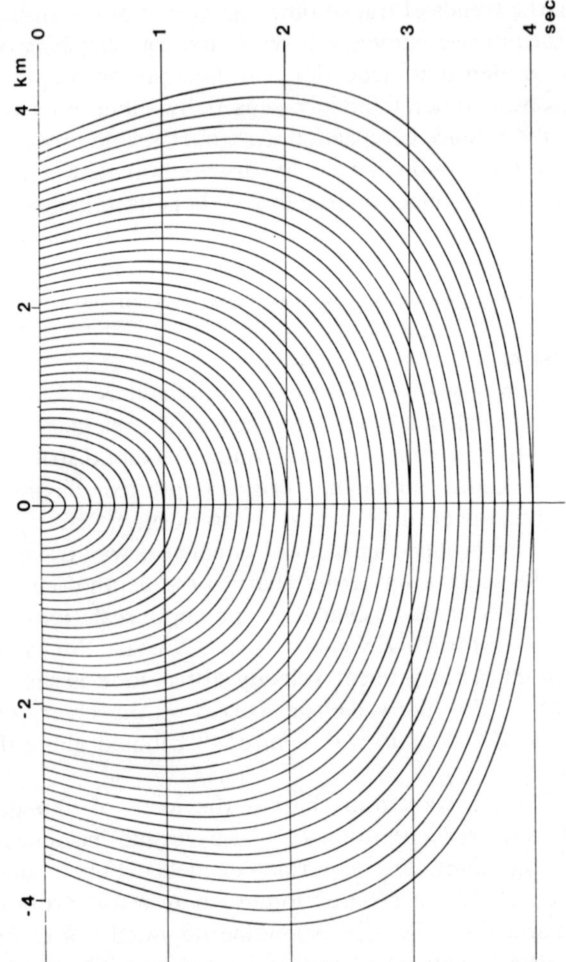

FIG. 7.8. t–x reflection wavefront chart for linear velocity distribution.

wavefront elements along a descending ray will generally increase with travel time up to the point where the ray reaches its maximum depth of penetration. This implies that where the time gradient along an unmigrated reflector is uniform, its migrated equivalent will be convex in depth and time domains. Similarly, when a sequence of unmigrated horizons is parallel, its migrated depth or time image will diverge in the downdip direction. An unmigrated reflector sequence which converges in the downdip direction may therefore correspond to a series of parallel layer interfaces.

Diffraction curve charts

A diffraction curve chart is composed of a set of diffraction t–x curves from a vertical array of diffraction point sources. Diffraction times are two-way times along raypaths which connect diffraction sources with zero-offset recording points along the X-axis of the chart. After plotting diffraction sources along its time axis, diffraction t–x curves can be constructed with the aid of an appropriate TX wavefront chart. Fig. 7.9 is an example of a diffraction curve chart for a linear velocity distribution.

When the velocity increases with depth there will for each diffraction source be a corresponding diffraction path which strikes it in a horizontal direction. These special raypaths can also be regarded as reflection paths at normal incidence to a vertical plane through the axis of the diffraction curve chart and at right angles to it. This leads to the property that the envelope of a diffraction curve chart represents the normal incidence reflection response of a vertical (fault) plane; it therefore defines, at each time level, the upper limit of the range of possible reflection time gradients for a given velocity distribution.

Two-dimensional migration

The two-dimensional migration of normal incidence reflection information along dip profiles may be carried out either in the depth domain by ray tracing, or in the time domain with the aid of wavefront or diffraction curve charts. The first of these two equivalent procedures is more practical during horizon migration (see Section 7.3). The second is directly related to signal migration principles and will be discussed here in some detail.

With reference to Fig. 7.10, let D be a hypothetical diffraction point source, coincident with the migrated reflection point R situated at a wavefront curve of time t. The corresponding unmigrated reflector

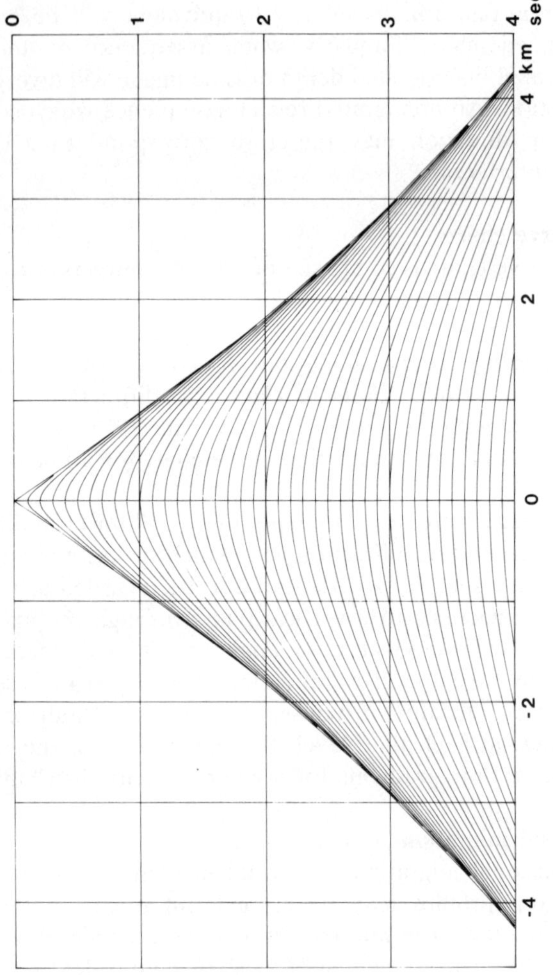

FIG. 7.9. t–x diffraction curve chart for linear velocity distribution.

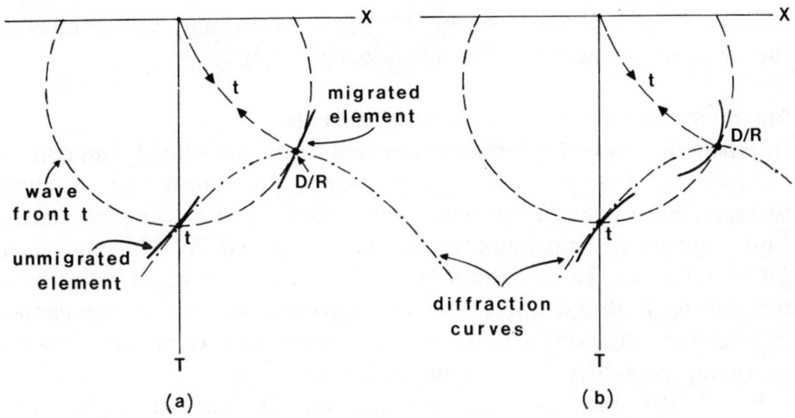

FIG. 7.10. Two-dimensional migration geometry in horizontal velocity distribution.

element passes through the axial point of the wavefront and is tangent to the diffraction curve from D. This follows from the consideration that at the origin of the wavefront chart reflection and diffraction time gradients are identical in view of common emergence angles of diffracted and reflected rays. A graphical migration can thus be carried out by determining the locus of culminations of diffraction curves tangent to an unmigrated reflection horizon when moving a properly oriented diffraction curve chart along the reflection profile. Hence, migration will transform the diffraction curve in Fig. 7.10 into a single point, the diffraction source at D.

From the relationship between the migrated and unmigrated elements in Fig. 7.10 we may summarize that:

1. A migrated reflector is the envelope of a set of wavefront curves whose axial points are situated on the corresponding unmigrated reflector.
2. A migrated reflector is the locus of culminations of diffraction curves tangent to its unmigrated image.
3. Corresponding migrated and unmigrated reflection horizons diverge in the downdip direction and intersect at the outcrop of the reflecting interface at the seismic datum plane.

The above divergence property may follow from a comparison of unmigrated and migrated time dips, which for a ray parameter p are $2p$

and $2p/(1-p^2V_z^2)^{\frac{1}{2}}$ respectively, V_z being the velocity at the level of the migrated reflection point (see Example 7.4).

Maximum and minimum convexity concepts

In reflection work time minima are traditionally associated with convex or anticlinal features and time maxima with concave or synclinal features. From this convention a diffraction curve is a convex entity. The expression 'maximum convexity', proposed by Hagedoorn in 1954, refers to the condition that where an unmigrated reflector is tangent to a diffraction curve its convexity (or curvature) cannot exceed that of the diffraction curve. The latter is therefore also called a maximum convexity curve or m.a.c. curve.

Fig. 7.10(a) illustrates the condition that the validity of the maximum convexity rule is subject to the constraint that the concavity of the migrated reflector element should be less than that of the wavefront element at their point of contact. This relationship always holds for anticlinal or monoclinal geologic features. The concavity of synclinal elements may exceed that of a wavefront curve, however. In that case the corresponding unmigrated reflector element will be more convex than the relevant diffraction curve [Fig. 7.10(b)]; the latter can then be described as a curve of minimum convexity.

The demonstration of the above properties with regard to curvature of unmigrated reflector elements can be based on the Fermat principle. Note that the reflection raypaths in Figs 7.10(a) and (b) are minimum and maximum time paths respectively. Fig. 7.11 refers to the case of the Fermat minimum. Reflection times plotted at R_1' and R_2' are the two-way times along normal incidence rays S_1R_1 and S_2R_2. Ray S_2R_1 is a diffraction path; the diffraction time observed at S_2 is plotted at point D, along the diffraction curve which culminates at R_1. In consequence of the Fermat condition the reflection time along S_2R_2 must be smaller than the diffraction time along S_2R_1. This is true for any recording position near S_1, which implies that at R_1' the convexity of the unmigrated element $R_1'R_2'$ cannot exceed that of the diffraction curve. Similarly, for the maximum time condition in Fig. 7.10(b) it will be found that the convexity of the unmigrated reflector element will exceed that of the diffraction curve.

The foregoing considerations can be extended to the three-dimensional case by rotating wavefronts and diffraction curves about their vertical axes. It can then be concluded that the interpreted relief of reflection time surfaces should satisfy limiting conditions with regard

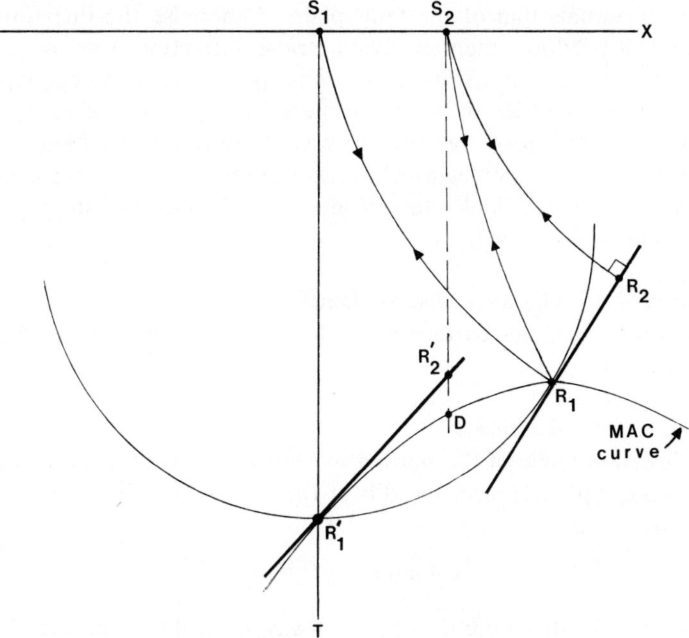

FIG. 7.11. Demonstration of maximum convexity property.

to curvature as well as reflection time gradient. Restrictions related to curvature are imposed by maximum or minimum convexity surfaces, and those pertaining to time gradients by the envelope of a chart of diffraction curves. Violation of these fundamental conditions will lead to anomalous migration results where reflection curvature is in conflict with the maximum or minimum convexity constraint, and to failure of migration during ray tracing where time gradients exceed their theoretical limits.

Maximum convexity curves registered on reflection profiles correspond to horizontal diffraction line sources oriented at right angles to profile planes. Where diffraction source and seismic traverse cross obliquely, however, the curvature of the diffraction event will be less than that of a theoretical m.a.c. curve. In the special case that line source and traverse are parallel, the diffraction event will be registered as a horizontal signal alignment which may be conformable to contiguous reflection patterns. Diffraction line sources along fault planes are horizontal when the layering is horizontal or where the strike of layer interfaces equals that of the fault plane. Otherwise the line source is inclined; it produces then an asymmetrical diffraction response, except where line source and traverse cross perpendicularly. Other types of diffraction geometries may be considered for curved line sources; an example related to a salt diapir was discussed in Section 7.2. In general, diffraction curves which deviate from theoretical m.a.c. curves transform into anticlinal artefacts after two-dimensional migration.

Example 7.4 *Migrated time gradient*
Show that for a ray parameter p the migrated time dip at depth z equals $2p/(1-p^2V_z^2)^{\frac{1}{2}}$.

Example 7.4 *Solution*
The direction angle of the normal incidence ray at the depth z follows from $\sin i_z = pV_z$. Hence, the dip of the migrated reflector element is given by

$$\mathrm{d}z/\mathrm{d}x = \tan i_z = pV_z/(1-p^2V_z^2)^{\frac{1}{2}}$$

From $\mathrm{d}z = \frac{1}{2}V_z\,\mathrm{d}t$, where $\mathrm{d}t$ is the two-way vertical time differential, we have after conversion into migrated time

$$\mathrm{d}t/\mathrm{d}x = 2p/(1-p^2V_z^2)^{\frac{1}{2}} = 2p/\cos i_z$$

Two-dimensional modelling

Fault bounded anticline
The depth model in Fig. 7.12 represents a rollover structure of the type frequently encountered in thick clastic sequences of Tertiary age. The normal fault is a major growth fault which originated by gravitational sliding of a normally compacted sand–shale section along the flank of an undercompacted and dominantly shaley formation in the form of a diapir. The undercompacted shales have a relatively low density and velocity and are in the deeper part of the upthrown fault block juxtaposed against the more sandy formation in the downthrown block. The model's reflection response is based on the velocity distribution $V_z = 1730 + 0{\cdot}57z$ m/s for the normally compacted intervals in up- and downthrown fault blocks, and a uniform velocity of 2260 m/s for the undercompacted shale unit. Fig. 7.13 shows the

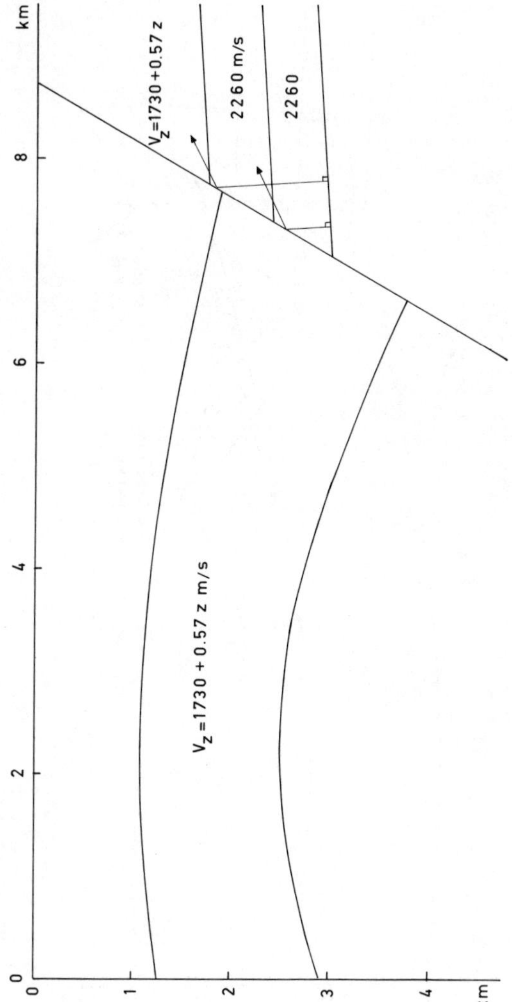

FIG. 7.12. Depth model of rollover structure bounded by growth fault.

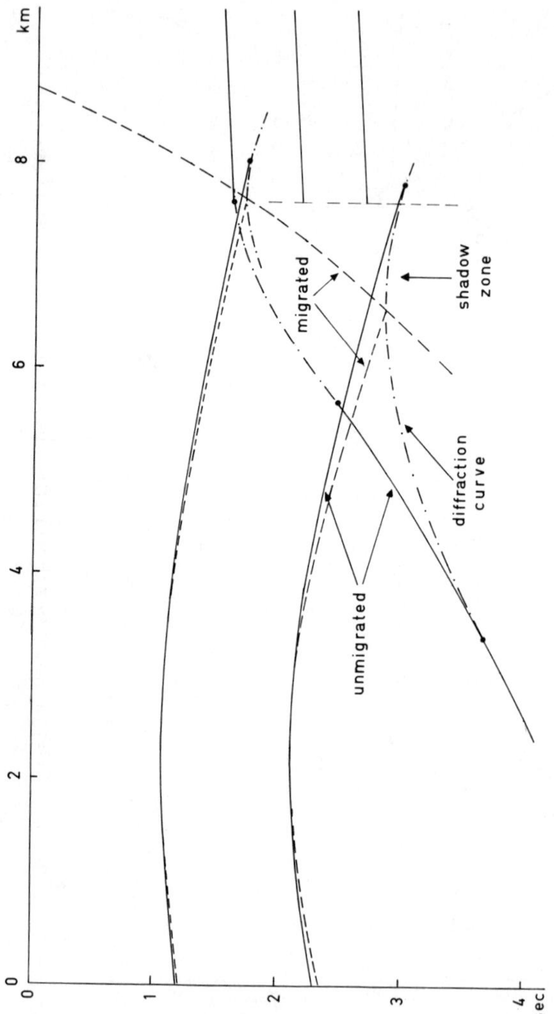

FIG. 7.13. Migrated and unmigrated time versions of depth model in Fig. 7.11.

migrated time version of the depth profile in Fig. 7.12 and its normal incidence reflection response. Principal points of interest are summarized as follows:

1. The relatively strong acoustic impedance contrast along the deeper part of the fault plane produces a fault plane reflection in the downthrown fault block.
2. Normal incidence rays from reflection horizons in the undercompacted shaley sequence in the upthrown fault block will strike the velocity discontinuity along the fault plane at an angle which exceeds the critical reflection angle. These rays will therefore undergo total reflection, as illustrated in Fig. 7.12. This causes the fade-out of reflection information from the region underneath the reflective part of the fault plane, commonly referred to as shadow zone.
3. From the migration geometry in Fig. 7.10, a diffraction curve from a diffraction source along the fault plane will be tangent to the fault plane reflection—or its theoretical extension—as well as to the reflections from horizons in up- and downthrown fault blocks which terminate at the diffraction source.

Practical examples of the seismic expression of fault-bounded roll-over structures are shown in profiles B, C and E. Profile B resembles the reflection response of the model in Fig. 7.12. The migrated section E exhibits the relationship between the fault's shadow zone and the migrated fault plane reflection. An interesting example of a well defined shadow zone is further shown along profile F. It is due to total reflection against a major reverse fault in a pre-Tertiary environment. The strong velocity contrast along the fault plane is evidenced by a distinct fault plane reflection in the upthrown fault block.

Faulted monocline

Fig. 7.14 shows the reflection response of a faulted monocline. The true position of the fault trace is displaced in the updip direction from its apparent position inferred from up- and downthrown reflector positions. In the absence of diffraction events there is here no firm basis for estimating the fault's true position, in contrast with the case of dip reversal in Fig. 7.13 where the fault could be traced with some confidence in the zone of intersecting reflection segments. From a formal point of view it may further be remarked that the apparent position of the fault in Fig. 7.14 is not identical to its unmigrated one; the latter corresponds to a recorded or imaginary fault plane reflection.

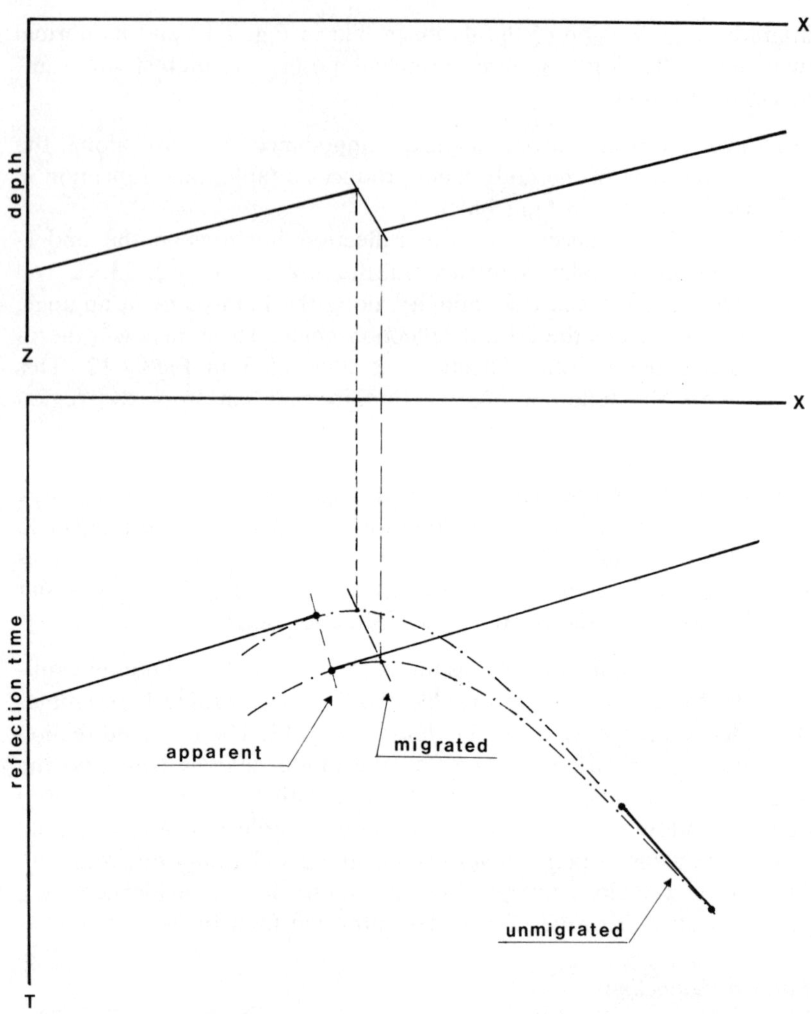

FIG. 7.14. Reflection response of faulted monocline.

Deep syncline

Fig. 7.15 shows the migrated time model of a synclinal feature characterized by the condition that the concavity of its bottom part exceeds that of contiguous wavefronts. Reflections from the shallower parts of the syncline intersect at the axial point of a wavefront curve tangent to both of its flanks. These flank reflections extend to unmigrated points

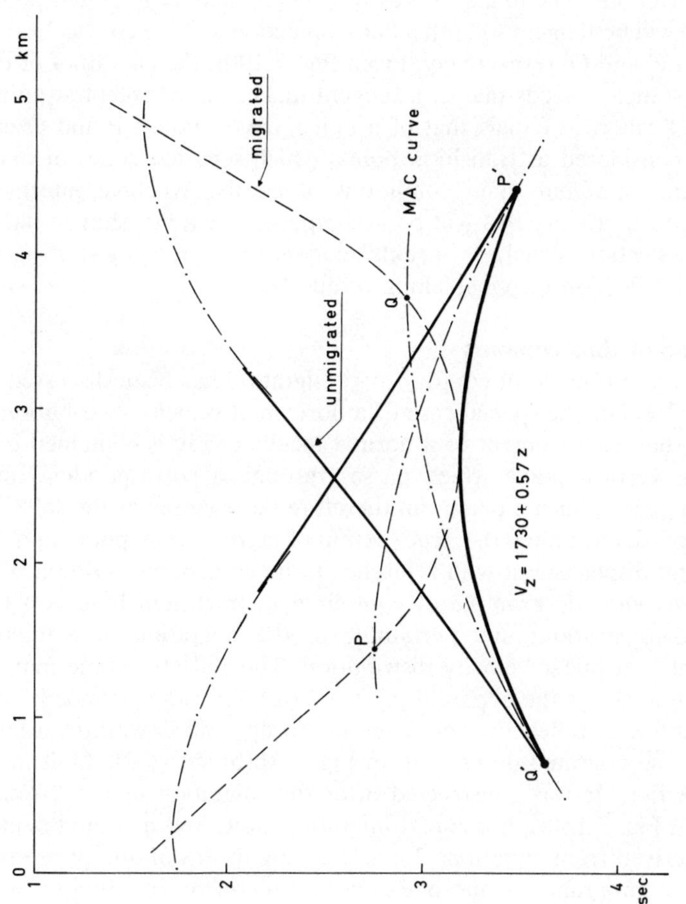

FIG. 7.15. Reflection response of deep syncline.

P′ and Q′. At the corresponding migrated points P and Q the curvature of the syncline will change from being less concave to being more concave than tangent wavefront curves. The response of the strongly curved part of the syncline between points P and Q corresponds to the convex branch connecting points P′ and Q′, where it merges with the flank reflections. The m.a.c. curves which culminate at migrated points P and Q will be tangent to both a flank reflection and the convex branch at points P′ and Q′ respectively. From Fig. 7.10(b) the curvature of the convex branch exceeds that of a tangent m.a.c. curve, except at points P′ and Q′ where it equals that of a m.a.c. curve. Points P and Q can also be considered as transition points which separate zones of maximum and minimum time conditions along the synclinal interface. From acoustic theory this will be accompanied by a 90° shift in phase of the reflection signal, as weakly evidenced by the registration of synclinal reflection response along profile N.

Migration of time contours

The general principle of contour map migration has been discussed in Section 7.3. For the special case of a horizontal velocity distribution it obtains that each element of a normal incidence ray is contained in a common vertical plane which passes through a corresponding time gradient vector. Such a plane can therefore be regarded as the solution plane for determining the true depth of a reflection point and its horizontal displacement with reference to its unmigrated position. Fig. 7.16 is a synthetic example of contour map migration based on the above considerations and pertaining to the migration of a faulted structure in a linear velocity distribution. The reflection time map in Fig. 7.16(a) shows the apparent positions of fault traces defined by the terminations of reflection time contours in up- and downthrown fault blocks. The migrated depth map in Fig. 7.16(b) shows the fault in its true position. It was constructed after the migration of the 14 data points in Fig. 7.16(a), horizontal migration shifts and migrated depths being derived from equations 3.6 and 3.7 by means of an appropriate calculator program. A more extensive treatment of contour map migration is included in Section 7.6.

Concluding problems

The next two examples are concerned with some three-dimensional aspects of reflection interpretation on normal incidence profiles.

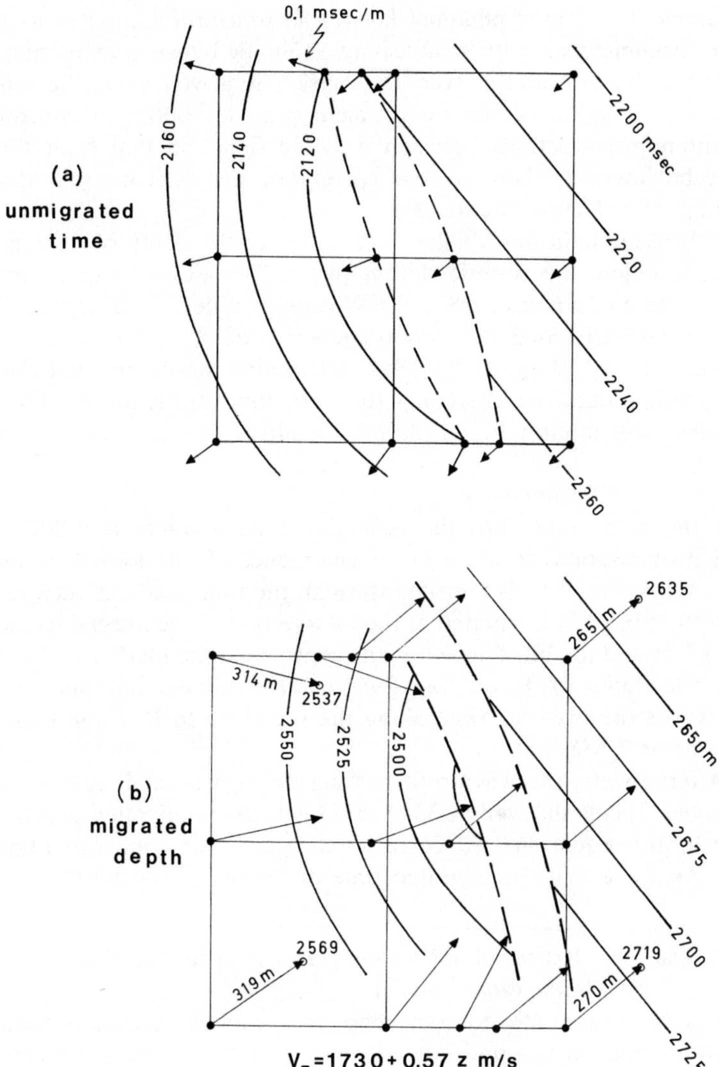

FIG. 7.16. Migration of reflection time contour map in horizontal velocity distribution.

Example 7.5 *Tie of phantom horizon to structural marker in well*
This example deals with establishing a time tie between an unmigrated phantom horizon and a structural marker in a well along the seismic traverse. Solution of the tie problem provides optimum control on phantom constructions between well locations, so that fault throws may be inferred where seismic correlation across faults fails due to change of reflection character.

A well log indicates a value of 3712 m for the depth of a structural marker below the seismic datum plane. The well is located at the intersection of a pair of NS and EW running reflection traverses. Time dips measured along the two reflection profiles in the zone of the marker are 0·150 and 0·260 s/km. Determine the unmigrated time to the marker interface, assuming that structural dip is locally uniform. Velocity distribution $V_z = 1730 + 0{\cdot}57z$ m/s.

Example 7.5 *Solution*
For the given time dips the reflection time gradient is 0·300 s/km, which corresponds to an angle of emergence of the normal incidence ray of 15°. Fig. 7.17 is a profile through the time gradient vector. The reflection point R is situated at the intersection of the normal incidence ray circle and the line connecting its centre with the marker in the well. For the depth of R we find 2966 m. The unmigrated time of the marker is the two-way time along the ray circle to R. From equation 3.7 this is 2·588 s.

Alternatively, the wavefront passing through point R may be constructed. It cuts the well at 3311 m. This is the unmigrated depth of R which, from equation 3.10, corresponds again to an unmigrated time of 2·588 s. Note that the migrated time of the marker is 2·803 s.

Example 7.6 *Reflection–diffraction response of faulted three-dimensional model*
Fig. 7.18 is a depth contour map of a faulted reflection horizon. Contour lines in up- and downthrown fault blocks are connected by contour lines along the fault plane. Construct the normal incidence reflection profile through the traverse indicated in the figure, showing

1. The reflection response of up- and downthrown horizons.
2. The fault plane reflection.
3. The diffraction response of the two line sources along the fault plane.

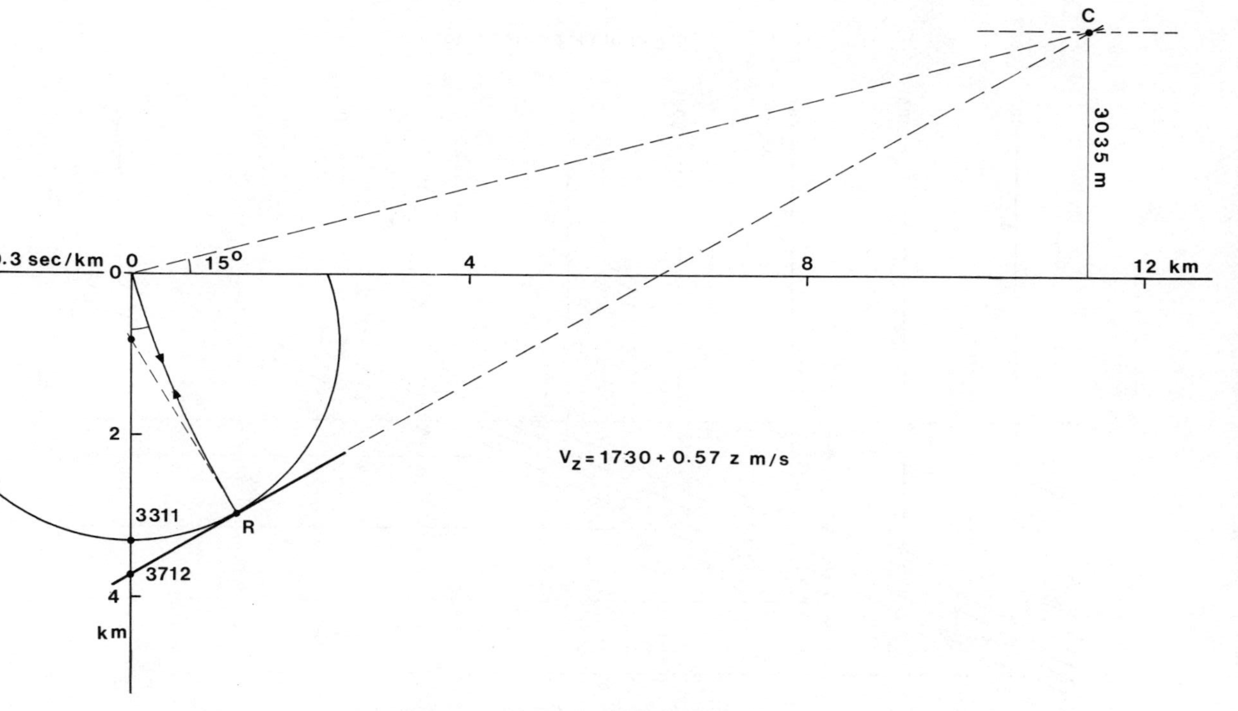

FIG. 7.17. Example 7.5 solution.

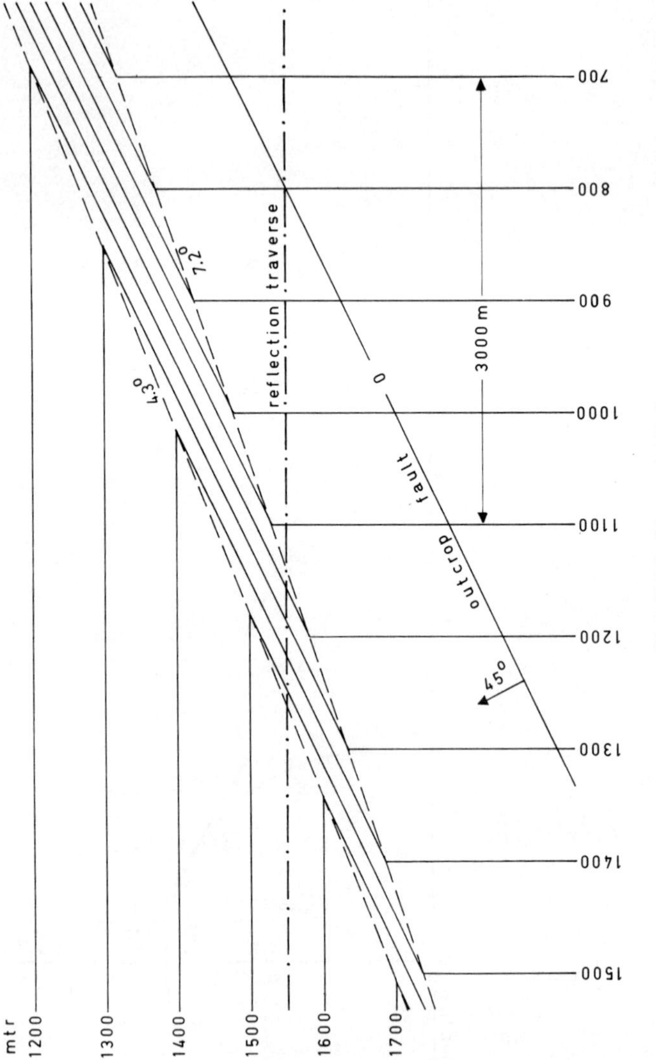

FIG. 7.18. Example 7.6.

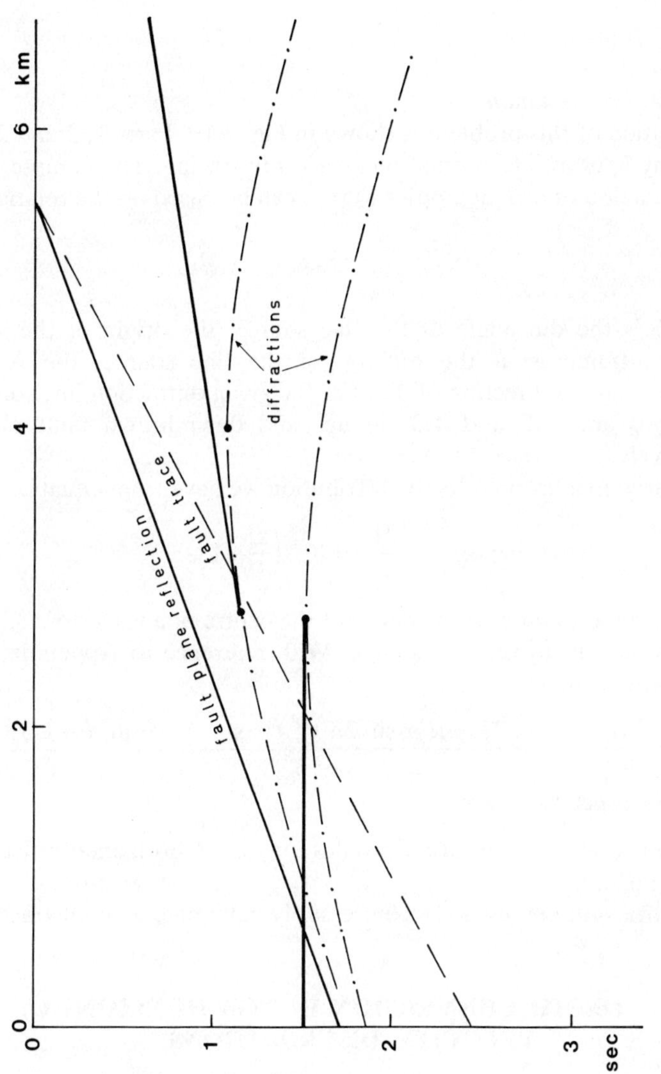

FIG. 7.19. Example 7.6 solution.

4. The fault trace, defined as the intersection of fault plane and the vertical plane of the reflection profile, converted into migrated time.

Velocity 2000 m/s.

Example 7.6 *Solution*
The solution of this problem is shown in Fig. 7.19. Parts 1, 2 and 3 are solved by appropriate normal incidence ray tracing. The computation of a diffraction time T at a point $P(x, y)$ can be based on the relation

$$T(x, y) = \frac{2}{V}(x^2 \sin^2 \delta + y^2)^{\frac{1}{2}}$$

where δ is the dip angle of the line source; the origin of the X–Y reference frame is at the outcrop of the line source, the X-axis extending into its direction of dip. For the given diffraction line sources dip angles are 7·2° and 4·3° in up- and downthrown fault blocks respectively.

Similarly, in a linear velocity distribution we have from equation 3.9

$$T(x, y) = \frac{2L}{V_0}\cosh^{-1}\left(\frac{z_c}{L}+1\right)$$

where $L = V_0/k$ and z_c is the depth of the centre of a wavefront sphere which is tangent to the line source. With reference to Appendix 3, it obtains for z_c that

$$z_c = \frac{-(L + \frac{1}{2}x \sin 2\delta) + [(\frac{1}{2}x \sin 2\delta + L)^2 + \sin^2 \delta(x^2 \sin^2 \delta + y^2)]^{\frac{1}{2}}}{\sin^2 \delta}$$

The main conclusions are

1. Fault traces on reflection profiles may cross horizontal reflection alignments.
2. Diffraction curves do not necessarily culminate at fault traces.

7.6 DEPTH CONVERSION IN NON-HORIZONTAL VELOCITY DISTRIBUTIONS

Introduction
Velocity distributions in sedimentary basins of pre-Tertiary origin usually consist of various components representing the velocity regimes

of geologic formations of widely different ages and compositions, separated by formation boundaries or unconformities. These outline the framework of a discontinuous velocity distribution, corresponding to significant acoustic impedance contrasts which can be traced by correlation of reflection character. Pertinent examples of such key horizons are the base Tertiary and base Lower Cretaceous unconformities, the base of the Upper Cretaceous, the top of the Triassic and the boundaries of the Zechstein formation in the Upper Permian. Corresponding reflection horizons are shown on profiles A, K, L, M, O and P from Western Europe.

Velocity distributions of the type described above are often not horizontal but may be disturbed by folding, faulting and diapiric intrusions. When a velocity unit is mainly composed of salt its velocity is nearly uniform and of the order of 4500 m/s. Otherwise intra-formation velocities may vary in the lateral direction and, discretely or continuously, along the formation interval. Continuous distributions are typical, for instance, of the Jurassic and Lower Cretaceous shales, and discrete distributions of the Triassic sequence consisting of layers of evaporites, sandstones and shales.

Depth conversions in discrete velocity distributions are commonly based on the interval velocities of the major velocity units. Establishing dependable estimates with regard to their spatial variations is one of the main interpretational problems in regions remote from well control. Other complicating factors are stacking difficulties mentioned in Section 7.1, limitations of a vertical depth conversion system and the complexity of map migration procedures.

Diffraction geometry

Fig. 7.20 shows the relationship between zero-offset reflection and diffraction geometries in a non-horizontal velocity distribution. Points R and R′ represent the migrated positions of a reflection point in *ZX* and *TX* domains respectively. A diffraction source at R will produce a diffraction curve which culminates where a diffraction raypath emerges vertically. The culmination C of the diffraction curve is in general displaced with reference to reflection point R′ in the horizontal as well as in the vertical directions. Note that from the Fermat principle the migrated time at R′ exceeds the diffraction time at recording point B, vertically above reflection point R.

Ray RA is at normal incidence to a reflector element at point R. In analogy with Fig. 7.10 the corresponding unmigrated reflector element

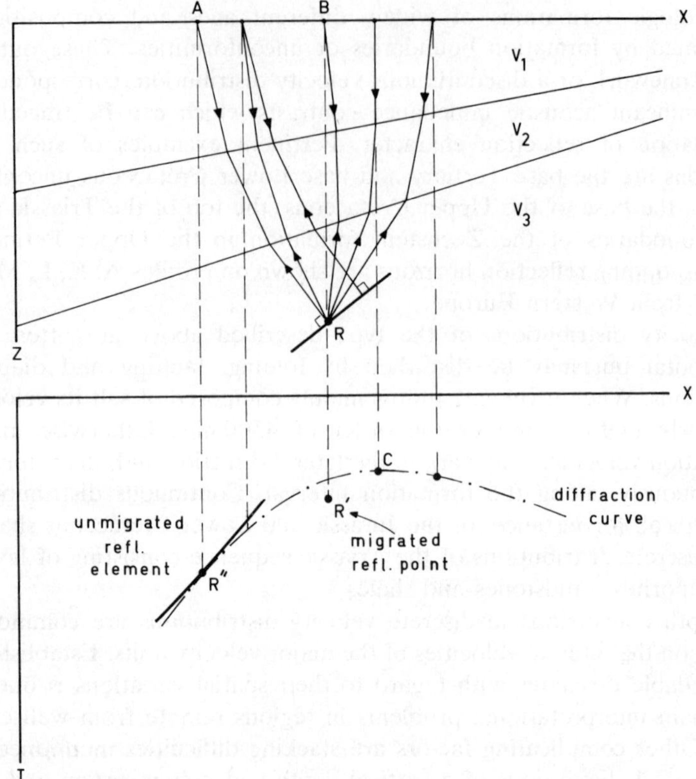

FIG. 7.20. Diffraction and migration geometry in non-horizontal velocity distribution.

will be tangent to the diffraction curve from R; the point of tangency R″ is the unmigrated reflection point, plotted vertically below recording position A. For reasons similar to the ones discussed in Section 7.5 we further have that at point R″ the convexity of the unmigrated element cannot exceed that of the diffraction curve from R.

From the above we may conclude that

1. On reflection profiles in non-horizontal velocity distributions diffraction curves from faults will in general not culminate at fault traces; the relative displacement depends upon depth and dip of the diffraction source, the attitudes of velocity interfaces and the magnitudes of velocity contrasts.

2. A migration process based on shifting unmigrated reflection points to m.a.c. curve culminations is restricted to horizontal velocity distributions.

Migration of reflection time contour maps

During routine reflection interpretation depth conversion is commonly performed by vertical scaling, which from Section 7.4 is a straightforward transformation of reflection time intervals into corresponding depth intervals, neglecting migration shifts. The three-dimensional migration of reflection horizons is considerably more cumbersome and therefore usually restricted to special cases where a more precise mapping of prospective features is required for the correct positioning of exploration or appraisal wells.

The preliminary phase of three-dimensional horizon migration is the organization of the input data to an appropriate map migration program based on three-dimensional ray tracing. This involves the compilation of velocity information and the construction of reflection time contour maps of the target horizon and of all significant velocity interfaces overlying it. These maps will then be sampled, establishing reflection times and reflection time gradients at the grid points of a square sampling grid. Such a sampling procedure may be carried out for instance by digitizing the intersections of contour lines and grid lines and fitting second or higher degree grid profiles to the digitized time values near grid points. From Section 7.3 reflection time gradients at grid points can then be expressed in terms of their components along grid lines by differentiation. The above approach can easily be adapted to situations whereby contour lines are interrupted by faults (Kleyn, 1977).

The migration of a sequence of layer interfaces is in essence an iterative process, that of the nth velocity interface involving the tracing of its normal incidence ray system through the $(n-1)$ overlying interfaces defined during preceding migration steps. After contouring and digitizing an intermediate migrated horizon, its depths and depth gradient components at the grid points of the sampling grid can be determined through a procedure identical to that applied to reflection times. This information will then be added to the input data required for the migration of deeper reflection horizons.

Three-dimensional ray tracing

The ray tracing algorithm described below is based on an approximation of the relief of migrated velocity interfaces by systems of plane elements, defined by depths and dip components at the grid points of the sampling grid. A normal incidence ray will be referred to a right-handed XYZ coordinate frame having its origin at the point of departure of the ray, the X- and Y-axis extending along grid lines. From Section 3.5 its initial direction is defined by the time gradient vector and the near-surface velocity at a relevant grid point. With reference to Fig. 7.21 let us assume that a ray descending from a grid

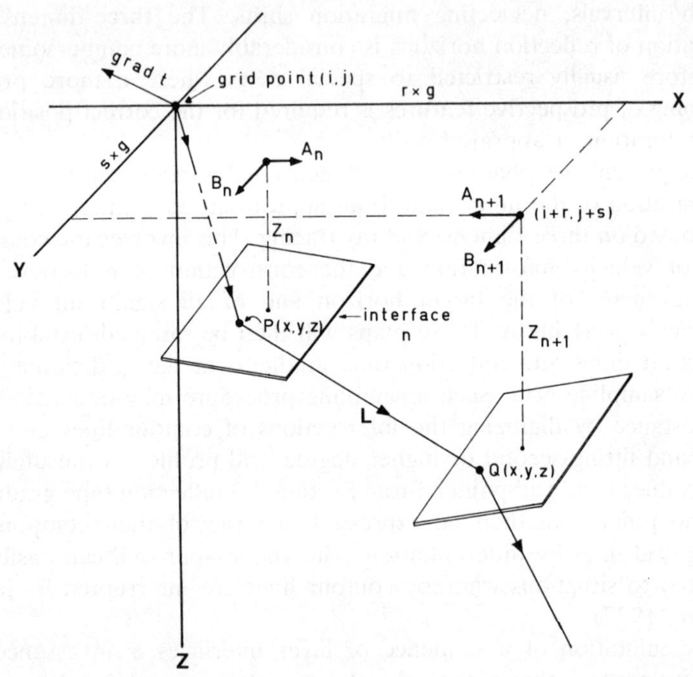

Fig. 7.21. Three-dimensional migration by ray tracing in non-horizontal velocity distribution.

point (i, j) has been traced to a point $P(x, y, z)$ located on an element of velocity interface n with dip components A_n and B_n in the X and Y directions. The transmitted ray vector (t_1, t_2, t_3) at point P can then be determined by substitution of the incident ray vector into the ray tracing equation of Section 3.5, along with the velocities in incident and transmitting media and the upward pointing normal vector at P, defined by its direction cosines

$$n_1 = A_n/(1+A_n^2+B_n^2)^{\frac{1}{2}}$$

$$n_2 = B_n/(1+A_n^2+B_n^2)^{\frac{1}{2}}$$

$$\text{and} \quad n_3 = -1/(1+A_n^2+B_n^2)^{\frac{1}{2}}$$

The next problem is the determination of the point of intersection $Q(x, y, z)$ of the transmitted ray and velocity interface $(n+1)$. Let us suppose that the ray tracing program finds point Q to be located at a velocity interface element at grid point $(i+r, j+s)$, where it is defined by a depth Z_{n+1} and dip components A_{n+1} and B_{n+1}. Its equation with reference to grid point (i, j) can then be written in the form

$$z = A_{n+1}x + B_{n+1}y + Z'_{n+1}$$

where

$$Z'_{n+1} = -(A_{n+1}r + B_{n+1}s)g + Z_{n+1}$$

and g is the spacing between grid points.

From the above it can easily be ascertained that the length L of the ray segment PQ is given by

$$L = \frac{A_{n+1}x_P + B_{n+1}y_P + Z'_{n+1} - z_P}{t_3 - t_1 A_{n+1} - t_2 B_{n+1}}$$

so that for the coordinates of point Q we find

$$x_Q = x_P + t_1 L$$

$$y_Q = y_P + t_2 L$$

$$\text{and} \quad z_Q = z_P + t_3 L$$

After determining the transmitted ray vector at point Q, the ray tracing continues until the accumulated two-way travel time along the traced path is equal to the reflection time at the point of departure of the ray, its termination representing a migrated reflection point.

Analysis of migration results

After the migration of a reflection horizon each grid point of the sampling grid can be related to the lateral position of a corresponding migrated reflection point. The set of migrated reflection points defines thus a migrated sampling grid, linked to the original sampling grid by horizontal displacement vectors. This will involve both a translation and a distortion of grid cells, depending on structural complexity. We may further distinguish between normal and anomalous grid cell distortions. For continuous reflection time surfaces a legitimate grid distortion requires that grid cell angles should not change by more than 90° during the migration. Otherwise the distortion is anomalous and indicative of discrepant input data to the migration program. These are invariably associated with violations of the maximum convexity constraint, which may result from one of the following conditions or a combination thereof:

1. Misjudgements with regard to the variation of time contour spacings in regions which lack detailed reflection coverage.
2. An incorrect assessment of interval velocity distributions.
3. Deviations of the positions of stacked signals from their theoretical normal incidence positions along horizons underneath complex velocity overburdens.

A typical example of an unacceptable distortion of a migrated sampling grid is shown in Fig. 7.22.

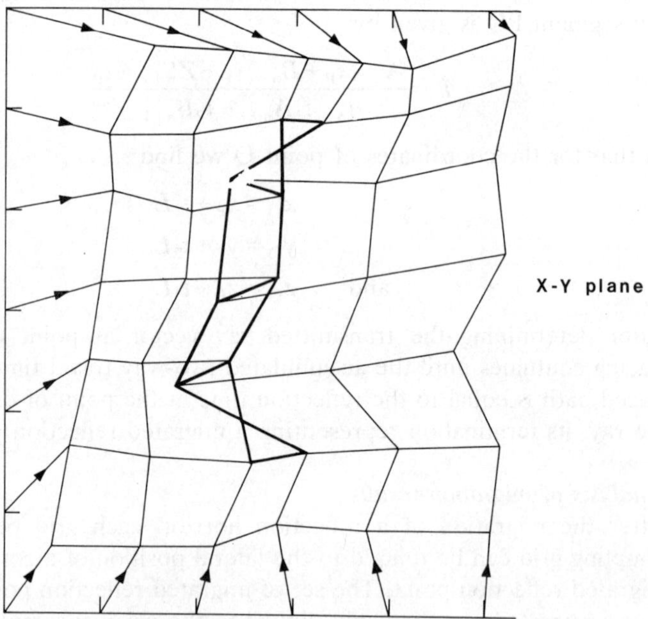

FIG. 7.22. Example of anomalous distortion of migrated sampling grid after contour map migration.

Map migration example
Fig. 7.23 illustrates the migration of a reflection time contour map traversed by a minor fault [Fig. 7.23(a)]. The overburden velocity distribution consists of two layers of uniform velocities, separated by the velocity interface shown in Fig. 7.23(b). The migrated depth map in Fig. 7.23(c) was obtained after the migration of the 16 grid points of a 300 by 300 m sampling grid defined by four rows and four columns, the migrated fault trace occupying an intermediate position between migrated grid points belonging to up- and downthrown fault blocks. The depth map manifests a potential fault trap which is not immediately evident from the unmigrated time map.

Example 7.7 *Horizon migration in non-horizontal velocity distribution*
Fig. 7.24 is a reflection time profile extending in the direction of subsurface dip and covering a faulted horizon underlying two velocity interfaces. Construct

1. The migrated depth section by normal incidence ray tracing.
2. The unmigrated depth section by vertical scale conversion.

Example 7.7 *Solution*
The migrated depth section in Fig. 7.25 is obtained by the successive migration of the two velocity interfaces and the target reflector. Emergence angles of normal incidence rays determined in accordance with the procedure of Example 7.3 are 0°, 11° and 1·5° for the first second and third horizon respectively. These rays are traced until two-way travel times equal the relevant reflection times at their points of departure. Note the lateral shift of the true position of the fault from its apparent position along the time section and the unmigrated depth section.

Vertical depth conversion
Vertical depth conversion of unmigrated reflection information is usually directly performed on reflection profiles after the digitization of time horizons. The computed depth values are subsequently plotted on situation maps and contoured to produce unmigrated depth maps. Such a procedure deviates from basic principles of seismic ray

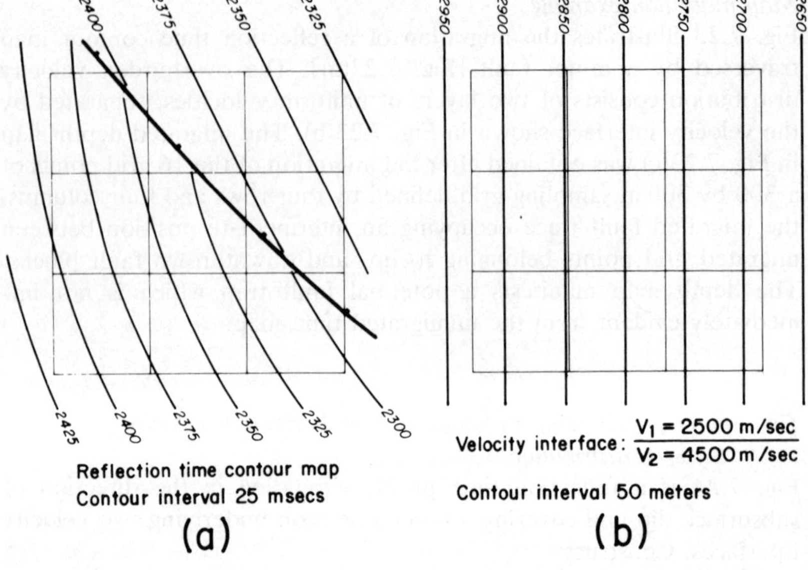

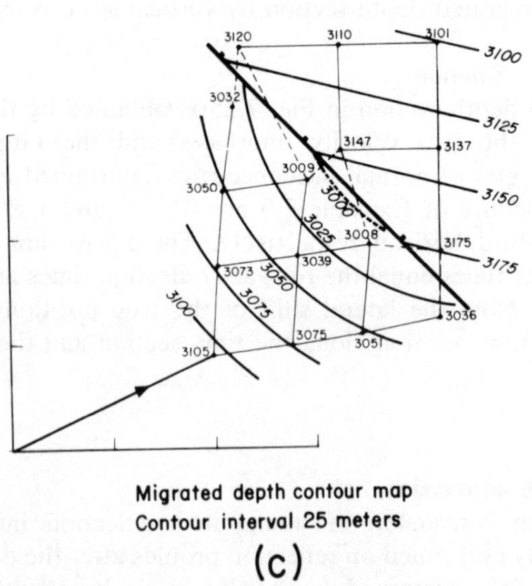

FIG. 7.23. Migration of reflection time contour map in non-horizontal velocity distribution.

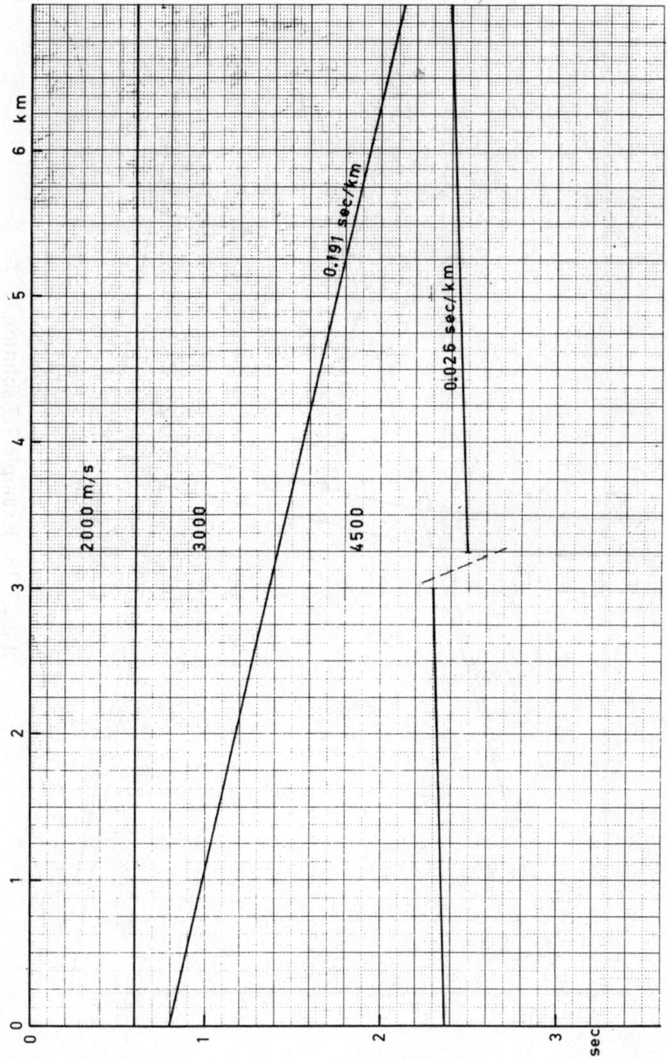

FIG. 7.24. Example 7.7.

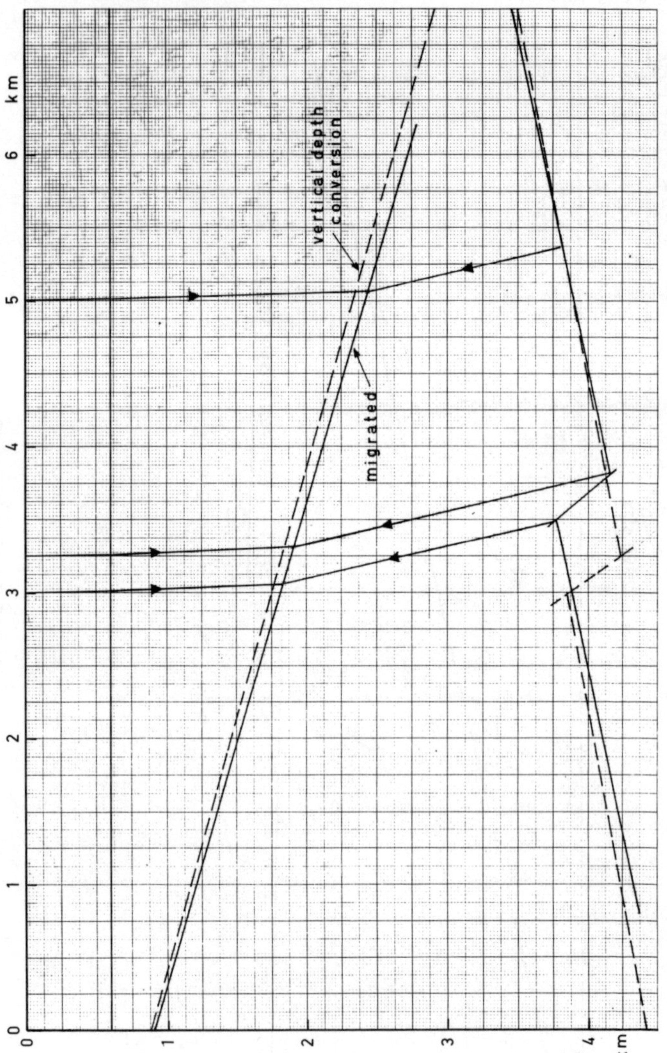

FIG. 7.25. Example 7.7 solution.

geometry but may lead to acceptable results in regions of low to moderate complexity, other merits being speed and simplicity. Depth domain interpretation through vertical depth conversion becomes unsatisfactory or impossible, however, under the following conditions

1. Crossover of reflection events from a single interface due to rapid changes of dip along it, or to a high relief velocity overburden (see Fig. 7.28).
2. Crossover of reflection events from a pair of interfaces with different inclinations. A typical example is reflection profile M: reflection events from the flank of a dolomite feature in the Zechstein formation arrive after the registration of the base Zechstein reflection.
3. Displacement of the velocity overburden along a major low angle fault, as exemplified on profile P.
4. Significant horizontal migration shifts of prospective features such as fault traps and pinch-outs underneath severely dipping velocity interfaces (see Figs 7.23 and 7.25).

It should further be borne in mind that a vertical depth conversion system cannot distinguish invalid data sets, incompatible with theoretical limits of reflection time gradient and reflection curvature.

For the case of a pair of continuous reflection horizons and minimum time conditions we shall analyse the errors inherent in vertical depth conversion with reference to Fig. 7.26. Let T_1 and T_2 be the normal incidence reflection times from horizons 1 and 2, registered at a point A along a reflection traverse. Corresponding migrated times measured along the vertical through A are T_1' and T_2' respectively. The unmigrated depth Z_a to the second horizon follows from

$$Z_a = \tfrac{1}{2}V_1T_1 + \tfrac{1}{2}V_2(T_2 - T_1)$$

whereas for the true depth we have

$$Z = \tfrac{1}{2}V_1T_1' + \tfrac{1}{2}V_2(T_2' - T_1')$$

Defining the depth error E by $E = Z_a - Z$, it then obtains that

$$E = \tfrac{1}{2}(V_1 - V_2)(T_1 - T_1') + \tfrac{1}{2}V_2(T_2 - T_2')$$

Since from the Fermat principle $T_1 < T_1'$ and $T_2 < T_2'$ the depth error will be negative when $V_1 > V_2$, or when reflector 1 is horizontal. Otherwise it may either be positive, negative or zero, depending on the type of velocity contrast and reflection time values, in accordance with

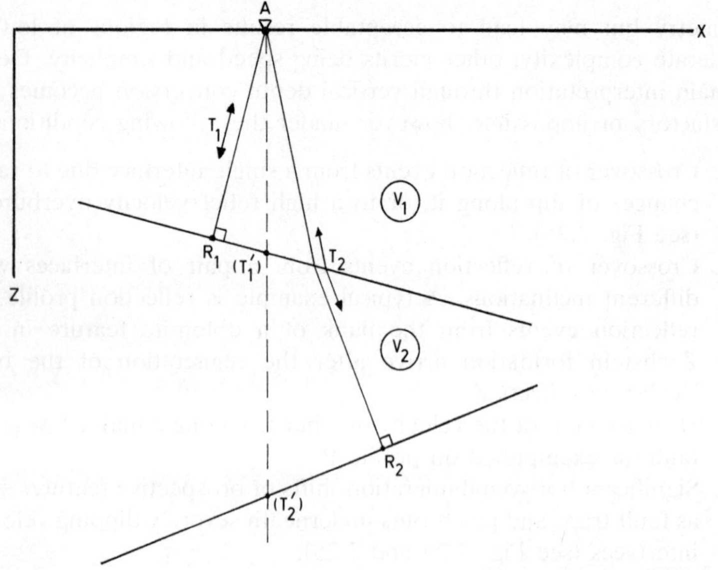

FIG. 7.26. Analysis of vertical depth conversion.

the results of Example 7.7 (Fig. 7.25). We further note that when on the time section in Fig. 7.26 the shallow reflector is placed in its migrated position, simulating a low angle fault trace, the depth error will be represented by the negative quantity $\frac{1}{2}V_2(T_2-T_2')$. In general it obtains for n reflectors that

$$E=\tfrac{1}{2}\sum_{1}^{n-1}(V_i-V_{i-1})(T_i-T_i')+\tfrac{1}{2}V_n(T_n-T_n')$$

leading to conclusions similar to the ones which apply to the two-layer case.

Vertical depth conversion is further illustrated in Figs 7.27 and 7.28. Fig. 7.27 is a cross-section through an anticline underneath an inclined velocity interface. Dips are exaggerated for clarity. On the depth profile migrated and unmigrated anticlinal arches intersect. Depth errors are positive along the crest of the anticline and negative along its flanks. Note that there is only a small offset of the anticline's apparent culmination with reference to its true position; the relative shift is in the downdip direction of the velocity interface.

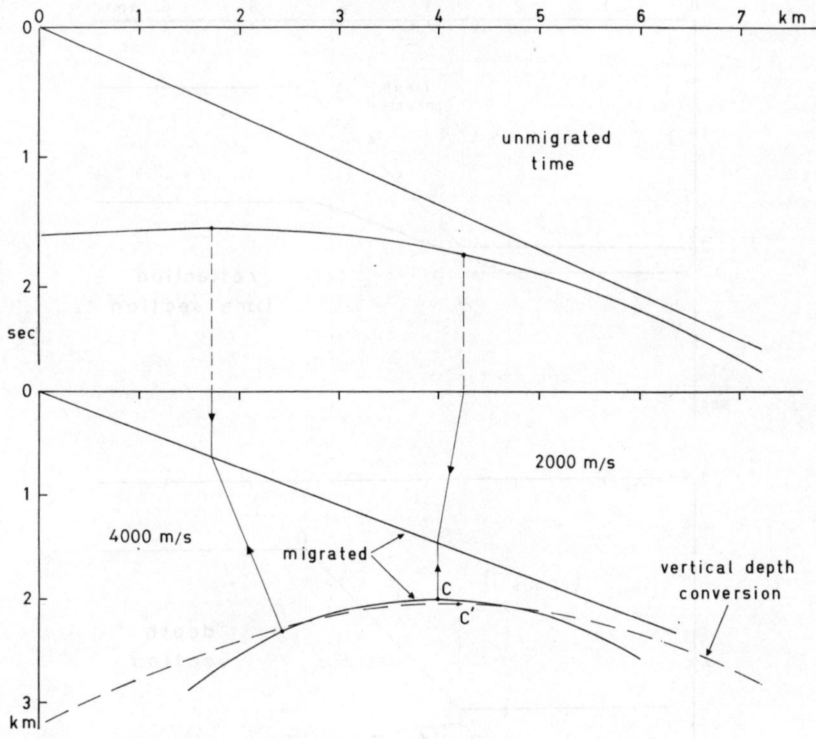

FIG. 7.27. Vertical depth conversion of anticlinal feature.

Fig. 7.28 exemplifies complicated velocity structure overlying a horizontal reflector. The steep flank of the velocity interface may represent a major fault or the face of a salt diapir. During the time domain interpretation such type of features are normally shown in their estimated true positions, inferred from unmigrated reflection evidence. Depth errors in this case result from vertical conversion into depth of an unmigrated horizon underlying a 'migrated' velocity interface and from the mismatch of corresponding dip changes along the two time horizons. The above refers to normal incidence data. In practice additional complications may arise through smearing of stacked signal alignments below the disturbed region of the velocity interface (see reflection profile P, p. 247).

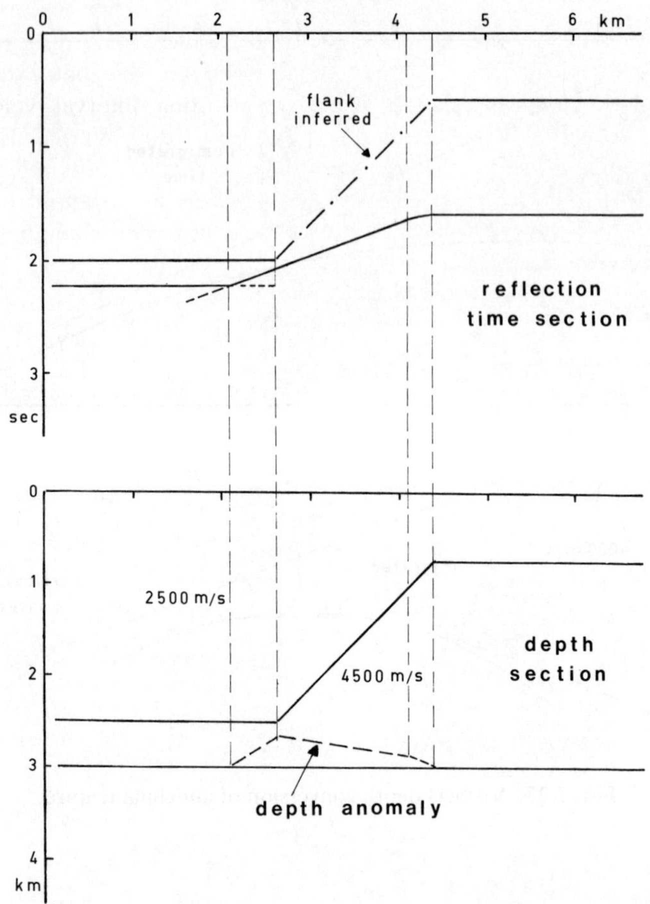

FIG. 7.28. False relief after vertical depth conversion through inferred migrated position of steeply inclined velocity interface.

Velocity control

Velocity control for the purpose of depth conversion of reflection information is primarily concerned with the optimum prediction of formation interval velocities in the regions between and encompassing a given set of velocity data points. These may be represented by well locations and seismic recording positions where interval velocities could with some confidence be derived from reflection moveouts.

As a first step in analysing lateral variations of interval velocities we may for each significant velocity unit investigate to what extent its interval velocity depends on depth by plotting interval velocities against the depths to the midpoint of the formation interval. Experience has shown that there exists sometimes a significant linear correlation between these two variables. Fig. 7.29 is an example for the Quaternary–Tertiary clastics and the Upper Cretaceous chalk based on well data from the Netherlands. Fig. 7.30 is a typical c.v.l. log of these formations. From a comparison of Figs 7.29 and 7.30 a distinction should generally be made between the velocity gradient of a distribution of instantaneous velocities measured along a formation interval and that of a distribution of interval velocities. Their mutual relationship is schematically illustrated in Fig. 7.31.

Let us suppose that there is in an ideal case a significant linear correlation between interval velocities and midpoint depths for all members of a sequence of velocity layers. Then, at a reflection observation point the interval velocity of each individual velocity unit can be predicted through the following procedure. When z_n and $\bar{z}_n$ are the depths to the top and the midpoint of a velocity layer n, T_{n1} and T_{n2} reflection times to its top and bottom and $\bar{V}_n(\bar{z}_n)$ the interval velocity to be estimated, it obtains with neglect of migration effects that

$$\bar{z}_n = z_n + 0{\cdot}25(T_{n2} - T_{n1})\bar{V}_n(\bar{z}_n)$$

Representing the regression of $\bar{V}_n$ on $\bar{z}_n$ by

$$\bar{V}_n(\bar{z}_n) = \bar{V}_n(0) + K\bar{z}_n$$

it follows that

$$\bar{V}_n(\bar{z}_n) = \frac{\bar{V}_n(0) + Kz_n}{1 - 0{\cdot}25K(T_{n2} - T_{n1})}$$

so that we have for the thickness Δz_n of layer n

$$\Delta z_n = \tfrac{1}{2}\bar{V}_n(\bar{z}_n)(T_{n2} - T_{n1})$$

Alternatively, we may regard a distribution of interval velocities as a pseudo-instantaneous velocity distribution along formation intervals, whose thicknesses can then be determined from equation 3.11. This will lead to near-identical results as can be demonstrated by means of Example 7.8.

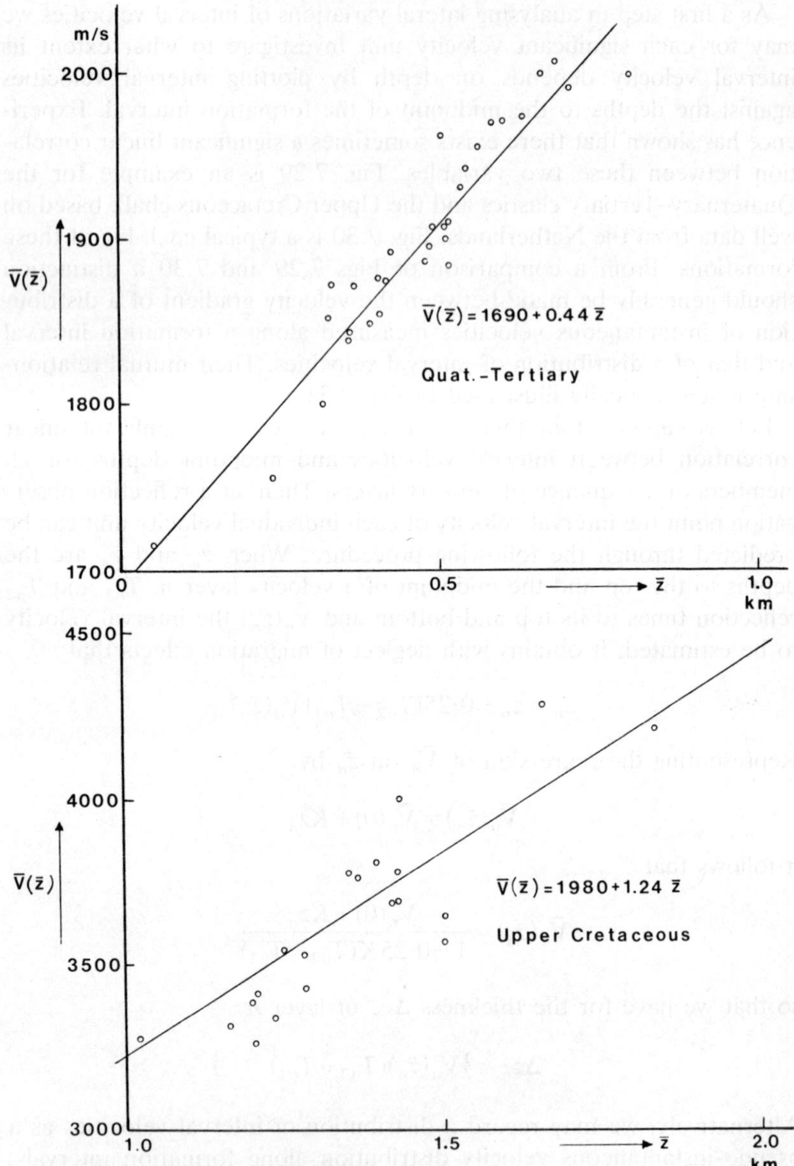

FIG. 7.29. Interval velocity versus depth plot for Quaternary–Tertiary clastics and Upper Cretaceous chalk based on well information in the Netherlands.

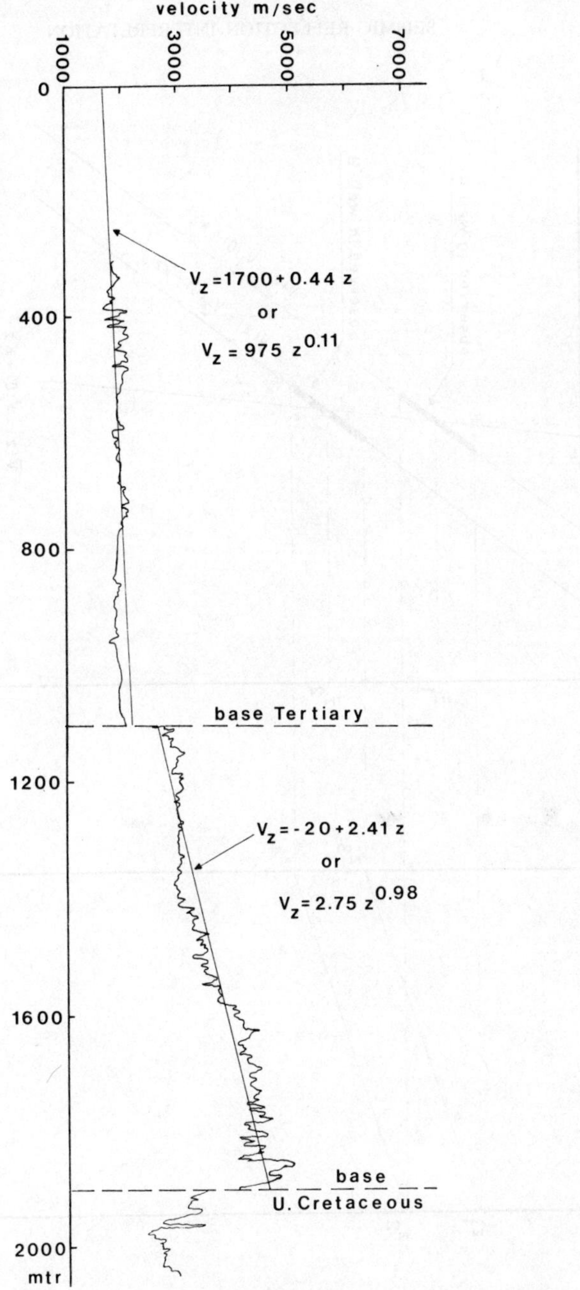

FIG. 7.30. Continuous velocity log for Quaternary–Tertiary and Upper Cretaceous formations in onshore well in the Netherlands.

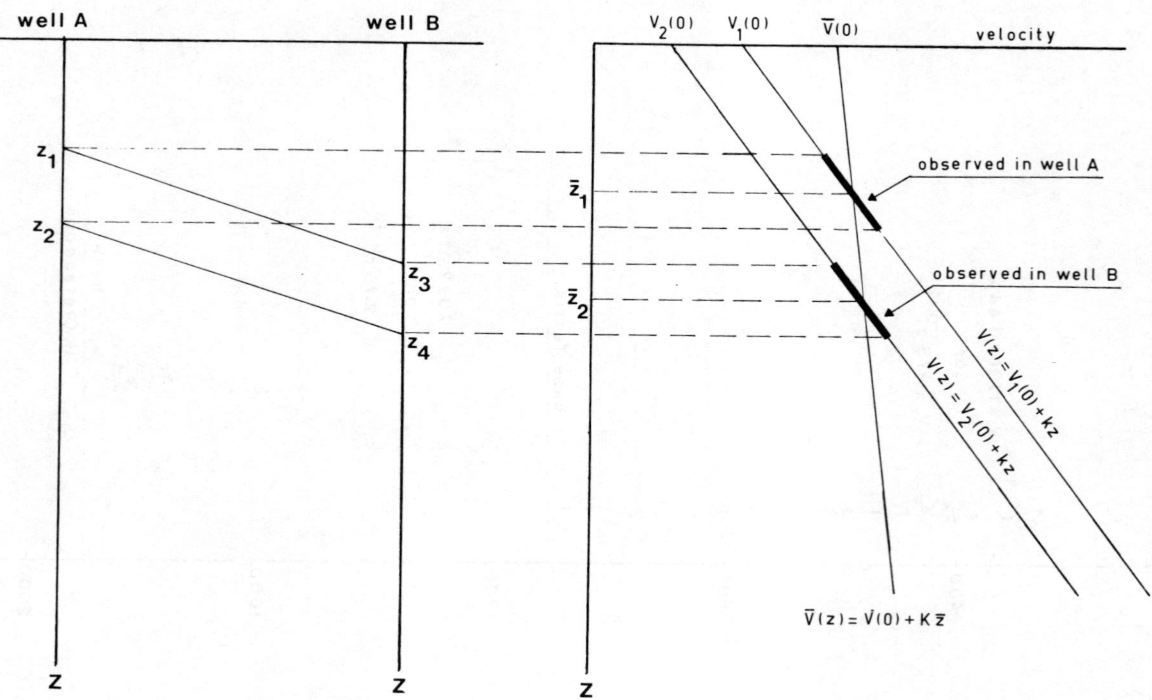

FIG. 7.31. Possible relationship between instantaneous and interval velocity distributions.

When there is a small to moderate scatter of interval velocities about a regression line, corresponding misties of reflection horizons at well locations may be eliminated by a judicious smoothing procedure, based for instance on the contouring of these discrepancies.

Where the interval velocity of a formation is mainly dependent on its original depth of burial there may be a poor correlation between interval velocities and present depth of burial in regions where tectonic inversion played an important role during the geological history of a depositional sequence. Representing for a specific velocity unit the inversion by $I(x, y)$, x and y being position coordinates, the relationship between its interval velocity and present depth of burial $\bar{z}_p$ can be written in the form

$$\bar{V}(x, y, \bar{z}_p) = \bar{V}_0 + K\bar{z}_p + KI(x, y)$$

where $\bar{V}_0$ is the initial velocity prior to the inversion. Assuming that in a restricted area the inversion is a linear function of x and y, a simple tilt, we have

$$\bar{V}(x, y, \bar{z}_p) = Px + Qy + K\bar{z}_p + R$$

P, Q, R and K are unknown coefficients to be determined by a *multiple regression analysis.*

The above may lead to a more accurate prognosis of interval velocities than can be obtained by a regression of $\bar{V}$ on $\bar{z}_p$ alone, as can be demonstrated through the following practical example. Fig. 7.32 is a plot of the interval velocities of a 300 m Lower Bunter shale section in the Triassic versus its midpoints depths from measurements in the five wells off the Dutch coast shown on the situation map in Fig. 7.33. The correlation between $\bar{V}_{LB}$ and $\bar{z}_p$ is poor; prediction errors for the Lower Bunter shale velocities at the well locations can be tabulated as follows:

Well	$\bar{V}_{LB}$ *observed* (m/s)	$\bar{V}_{LB}$ *predicted* (m/s)	*Pred. error* (m/s)
A	4 221	4 368	+147
B	3 624	3 747	+123
C	3 902	4 167	+265
D	4 563	4 397	−166
E	4 372	4 003	−369
			233 r.m.s.

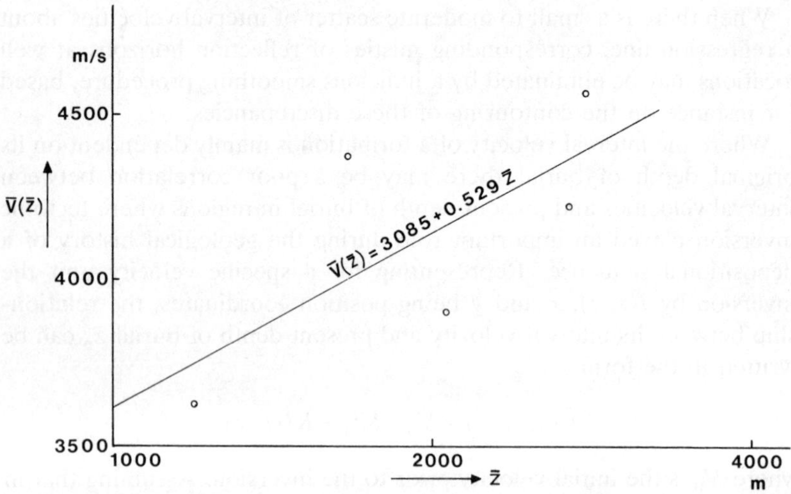

FIG. 7.32. Plot of interval velocities of Lower Bunter shale section versus midpoint depths observed in Netherlands offshore wells (ref. Fig. 7.33).

For the given data set a multiple regression of $\bar{V}_{\mathrm{LB}}$ on x, y and $\bar{z}_{\mathrm{p}}$ yields

$$\bar{V}_{\mathrm{LB}}(x, y, \bar{z}_{\mathrm{p}}) = -0{\cdot}025\,67x - 0{\cdot}031\,34y + 0{\cdot}486\,49\bar{z}_{\mathrm{p}} + 3060{\cdot}4 \quad \mathrm{m/s}$$

with x, y and $\bar{z}_{\mathrm{p}}$ in metres. From this relation an updated prognosis of Lower Bunter shale velocities is summarized as follows:

Well	$\bar{V}_{\mathrm{LB}}$ *observed* (m/s)	$\bar{V}_{\mathrm{LB}}$ *predicted* (m/s)	*Pred. error* (m/s)
A	4 221	4 240	+19
B	3 624	3 593	−31
C	3 902	3 927	+25
D	4 563	4 510	−53
E	4 372	4 396	+24
			33 r.m.s.

In general, when results of multiple regression analyses are significant the interval velocity of a layer n at a reflection observation point (x, y) may be predicted after substitution of the numerator in the foregoing expression for $\bar{V}_n(\bar{z}_n)$ by $\bar{V}_n(x, y, \bar{z}_{\mathrm{p}})$.

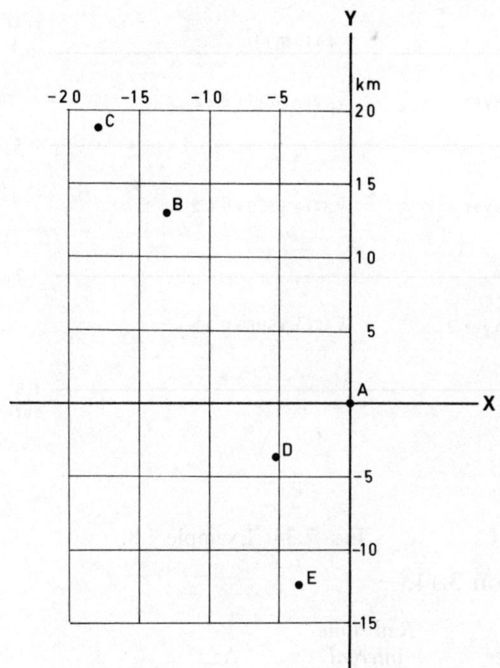

FIG. 7.33. Situation map for offshore wells referred to in Fig. 7.32.

Example 7.8 *Depth conversion*
Determine formation thicknesses for the data set in Fig. 7.34 from foregoing expressions for $\bar{V}_n(\bar{z}_n)$ and Δz_n, and from equation 3.11.

Example 7.8 *Solution*
From the expressions for $\bar{V}_n(\bar{z}_n)$ and Δz_n:

Layer	*Refl. time interval*	$\bar{V}_n(\bar{z}_n)$	Δz_n	z_n
1	0·5	1 641	410	0
2	0·7	3 402	1 191	410
3	0·6	4 168	1 250	1 601
4	—	—	—	2 851

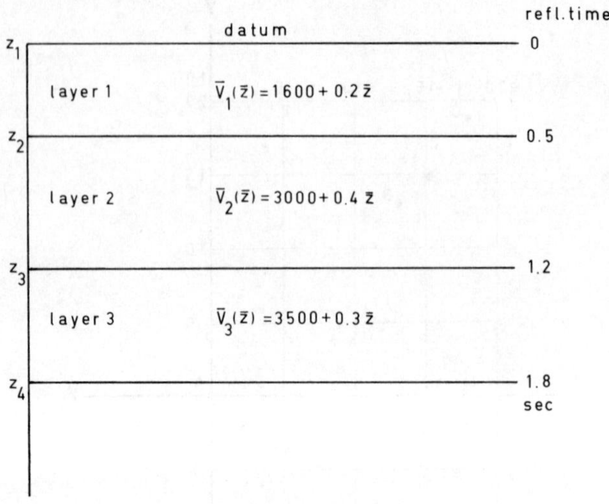

FIG. 7.34. Example 7.8.

From equation 3.11,

Layer	*Refl. time interval*	Δz_n	z_n	$\bar{V}'_n(0)$
1	0·5	410	0	1 600
2	0·7	1 189	410	3 164
3	0·6	1 249	1 600	3 980
4	—	—	2 849	—

In the tabulation above $\bar{V}'_n(0)$ represents an initial velocity value, to be substituted for V_0 in equation 3.11 for computing the thickness of layer *n*.

REFERENCES

Hagedoorn, J. G. (1954). A process of seismic reflection interpretation, *Geophysical Prospecting*, **2**(2).

Kleyn, A. H. (1977). On the migration of reflection time contour maps, *Geophysical Prospecting*, **25**(1).

BIBLIOGRAPHY

Weatherby, B. B. and Faust, L. Y. (1935). Influence of geological factors on longitudinal seismic velocities, *Bull AAPG*, **19**, 1–8.

West, S. S. (1950). Dependence of seismic wave velocity upon depth and lithology, *Geophysics*, **15**(4).
Faust, L. Y. (1951). Seismic velocity as a function of depth and geologic time, *Geophysics*, **16**(2).
Kaufman, H. (1953). Velocity functions in seismic prospecting, *Geophysics*, **18**(2).
Faust, L. Y. (1953). A velocity function including lithologic variation, *Geophysics*, **18**(2).
Hagedoorn, J. G. (1954). A process of seismic reflection interpretation, *Geophysical Prospecting*, **2**(2).
Musgrave, A. W. (1961). Wave front charts and three-dimensional migrations, *Geophysics*, **26,** 738–753.
Sattlegger, J. W. (1964). Series for three-dimensional migration in reflection seismic interpretation, *Geophysical Prospecting*, **12,** 115–134.
Sattlegger, A. W. (1969). Three-dimensional seismic depth computation using space-sampled velocity logs, *Geophysics*, **34,** 7–20.
Taner, M. T., Cook, E. E. and Neidell, N. S. (1970). Limitations of the seismic reflection method; lessons from computer simulations, *Geophysics*, **35**(4).
Paturet, D. (1971). Different methods of time–depth conversion with and without migration, *Geophysical Prospecting*, **19**(1).
Sorrells, G. G., Crowley, J. R. and Veith, K. F. (1971). Methods for computing ray paths in complex geological structures, *Bull. Seism. Soc. Am.*, **61,** 27–53.
Tucker, P. and Yorston, H. (1973). Pitfalls in seismic interpretation, SEG Monograph no. 2.
Dobecki, T. L. (1973). Three-dimensional seismic modeling for arbitrary velocity distributions, *Geophysical Prospecting*, **21,** 330–339.
Magara, K. (1976). Thickness of removed sedimentary rocks, paleopore pressure, and paleotemperature, southwestern part of Western Canada basin, *AAPG Bull.*, **60**(4).
Fitch, A. A. (1976). *Seismic Reflection Interpretation*, Gebrüder Borntraeger, Berlin–Stuttgart.
Kleyn, A. H. (1977). On the migration of reflection time contour maps, *Geophysical Prospecting*, **25**(1).
Gerritsma, P. H. A. (1977). Time-to-depth conversion in the presence of structure, *Geophysics*, **42**(4).
Anstey, N. A. (1977). *Seismic Interpretation: the Physical Aspects*, International Human Resources Development Corporation, Boston.
Crans, W., Mandl, G. and Haremboure, J. (1980). On the theory of growth faulting: a geomechanical delta model based on gravity sliding, *Journal of Petroleum Geology*, **2**(3).
Lang, W. H. (1980). Determination of prior depth of burial using interval transit time, *Oil & Gas Journal* (Jan. 28).
Sheriff, R. E. (1980). *Seismic Stratigraphy*, International Human Resources Development Corporation, Boston.
Acheson, C. H. (1981). Time–depth and velocity–depth relations in sedimentary basins—A study based on current investigation in the Arctic Islands and an interpretation of experience elsewhere, *Geophysics*, **46**(5).
Sattlegger, J. W., Rhode, J., Egbers, H., Dohr, G. P., Stiller, P. K. and

Echterhoff, J. A. (1981). INMOD—Two dimensional inverse modeling algorithm based on ray theory, *Geophysical Prospecting*, **29**(2).

Schilt, F. Steve, Kaufman, S. and Long, G. H. (1981). A three-dimensional study of seismic diffraction patterns from deep basement sources, *Geophysics*, **46** (December).

Zawisiak, R. L. and Smithson, S. B. (1981). Problems and interpretation of COCORP deep seismic reflection data, Wind River range, Wyoming, *Geophysics*, **46** (December).

Sattlegger, J. W. (1982). Migration of seismic interfaces, *Geophysical Prospecting*, **30**(1).

CHAPTER 8

Elements of Signal Migration Systems

8.1 INTRODUCTION TO POST-STACK MIGRATION

Early signal migration programs were based on the migration geometry in Fig. 7.10. Its application to digital migration is illustrated in Fig. 8.1 A set of closely spaced maximum convexity curves is moved along the unmigrated reflection time section in steps of one trace interval. After each step trace samples intercepted by the m.a.c. curves are shifted to their culminations and added to produce a so-called unweighted diffraction stack. This is referred to as two-dimensional migration by summation, or integration of signal samples along diffraction curves. From Fig. 8.2 such a process is equivalent to distributing signal samples along wavefront curves. Migrated events emerge along envelopes tangent to these wavefronts, signal samples elsewhere interfering destructively. Modern wave equation migration systems produce similar effects but incorporate fundamental acoustic principles neglected by the older migration methods. The unweighted diffraction stack suffers therefore amplitude and phase distortion and shows an imperfect cancellation of signal samples at wavefront intersections. The above is restricted to normal incidence reflection data and diffractions. Spurious events and random noise survive after their dispersion along wavefronts. These are then manifested as 'smiles' or 'rooster tails' in zones along migrated profiles with a low signal to noise ratio.

From Section 3.8 and in analogy with the t–x relation for c.m.p. gathers maximum convexity curves can be closely approximated by the hyperbolic expression

$$\tau_L^2 = \tau_0^2 + \frac{4L^2}{V_{\mathrm{rms}}^2(\tau_0)}$$

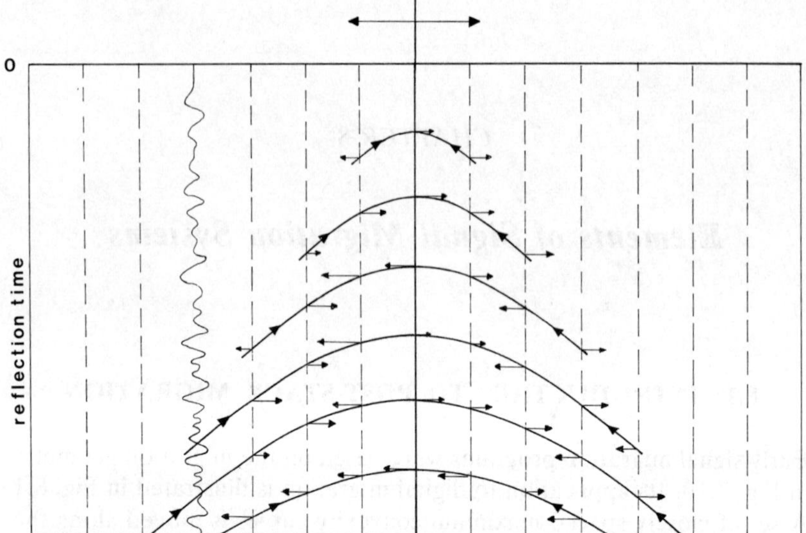

FIG. 8.1. Two-dimensional signal migration. Generation of unweighted diffraction stack by summation of signal samples along m.a.c. curves.

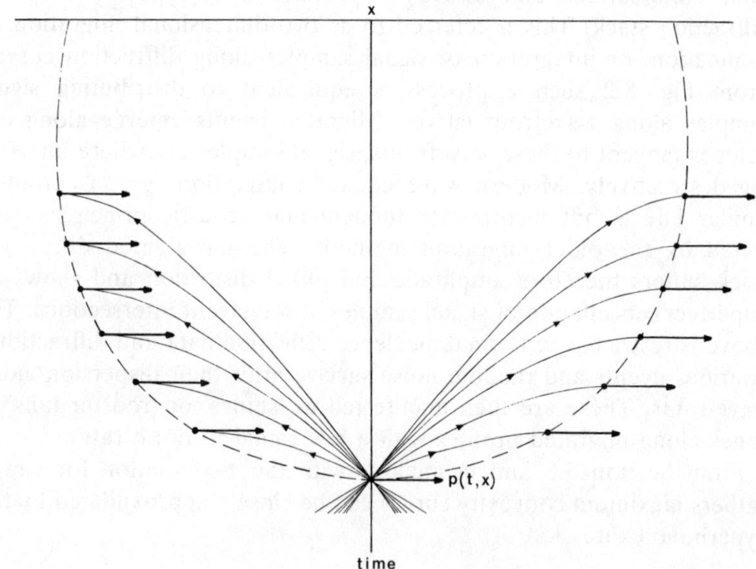

FIG. 8.2. Distribution of signal samples along wavefront curves during two-dimensional signal migration.

where L is the distance coordinate with reference to their central axes and τ the two-way diffraction time. V_{rms} may be estimated from stacking velocities by considering that for a reflector element with a dip δ we have $V_s \approx V_{rms}/\cos\delta$. The corresponding unmigrated time dip is approximated by $dt/dx = 2\sin\delta/V_{rms}$, so that after the elimination of δ we find

$$\frac{1}{V_{rms}^2} \approx \frac{1}{V_s^2} + \frac{1}{4}\left(\frac{dt}{dx}\right)^2$$

In mathematical descriptions of migration procedures we shall distinguish between a *migrated reflection time T* and a *diffraction time* τ. With reference to the position coordinates in Fig. 8.3 the unweighted

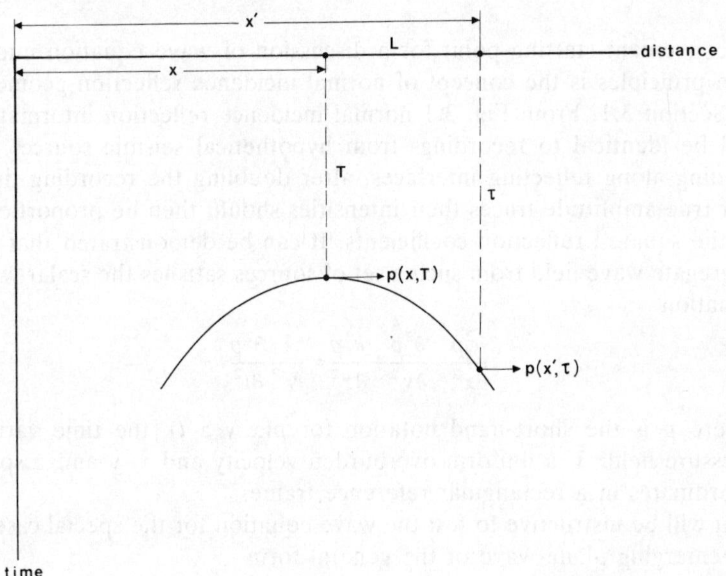

FIG. 8.3. System of symbols for mathematical description of two-dimensional signal migration.

diffraction stack can then be defined by the integral expression

$$p(x, T) = C\int_{-L_{max}}^{+L_{max}} p(x', \tau)\, dx'$$

with

$$\tau^2 = T^2 + \frac{4L^2}{V^2_{\mathrm{rms}}(T)}$$

C is a scale factor. Similarly, we have for areal reflection data

$$p(x, y, T) = C \int_{x'} \int_{y'} p(x', y', \tau)\,\mathrm{d}x'\,\mathrm{d}y'$$

with

$$L^2 = (x' - x)^2 + (y' - y)^2$$

8.2 WAVE EQUATION MIGRATION

A convenient starting-point for a discussion of wave equation migration principles is the concept of normal incidence reflection geometry of Section 3.1. From Fig. 3.1 normal incidence reflection information will be identical to recordings from hypothetical seismic sources extending along reflecting interfaces, after doubling the recording time. For true amplitude traces their intensities should then be proportional to the squared reflection coefficients. It can be demonstrated that the aggregate wave field from such a set of sources satisfies the scalar wave equation

$$\frac{\partial^2 p}{\partial x^2} + \frac{\partial^2 p}{\partial y^2} + \frac{\partial^2 p}{\partial z^2} = \frac{1}{V^2}\frac{\partial^2 p}{\partial t^2}$$

where p is the short-hand notation for $p(x, y, z, t)$, the time variant pressure field, V a uniform overburden velocity and x, y and z space coordinates in a rectangular reference frame.

It will be instructive to test the wave equation for the special case of an emerging plane wave of the general form

$$p\left(t - \frac{x}{V_x} - \frac{y}{V_y} - \frac{z}{V_z}\right)$$

V_x, V_y and V_z are apparent velocities in the directions of the X-, Y- and Z-axes, related to the true propagation velocity V by

$$1/V^2 = 1/V_x^2 + 1/V_y^2 + 1/V_z^2$$

When p'' is the second order derivative of p with reference to its bracketed argument we have after performing the differentiations

$$\frac{\partial^2 p}{\partial x^2} = \frac{p''}{V_x^2}$$

$$\frac{\partial^2 p}{\partial y^2} = \frac{p''}{V_y^2}$$

$$\frac{\partial^2 p}{\partial z^2} = \frac{p''}{V_z^2}$$

$$\frac{\partial^2 p}{\partial t^2} = p''$$

which fulfils the wave equation, as can be verified by substitution.

When the origin of the reference frame is at the earth's surface, $p(x, y, 0, t)$ represents a recorded normal incidence data set. The initial phase of wave equation migration is the determination of $p(x, y, z, t)$ by deriving a solution of the wave equation which satisfies the boundary condition $p(x, y, 0, t)$ and the additional constraint that $p(x, y, 0, t) = 0$ for t larger than a preset maximum value. In other words, by solving the wave equation we can predict a normal incidence reflection field in any fictive recording plane below the earth's surface. This is referred to as *downward continuation* or *downward extrapolation* of reflection information. Since there is no migration effect at the outcrops of reflecting interfaces it follows correspondingly that for any value of z a *migrated sample* occurs at zero time. Consequently, migrated samples are represented by $p(x, y, z, 0)$ and migrated time by $T = 2z/V$. Singling out migrated samples is called *imaging*. Fig. 8.4 shows a geometrical interpretation of a downward continuation and imaging process.

During seismic data processing, time domain solutions of the wave equation may be obtained by two alternative procedures. The first is a direct solution, employing a special numerical technique called the *finite difference method*. The second is the *Kirchoff summation approach*, based on the evaluation of the Kirchoff integral expression, which satisfies the wave equation and can be derived from it. In the present section we shall consider basic principles of both of these methods. For a discussion of details and related topics such as migration in the wave-number/frequency domain reference is made to the relevant bibliography.

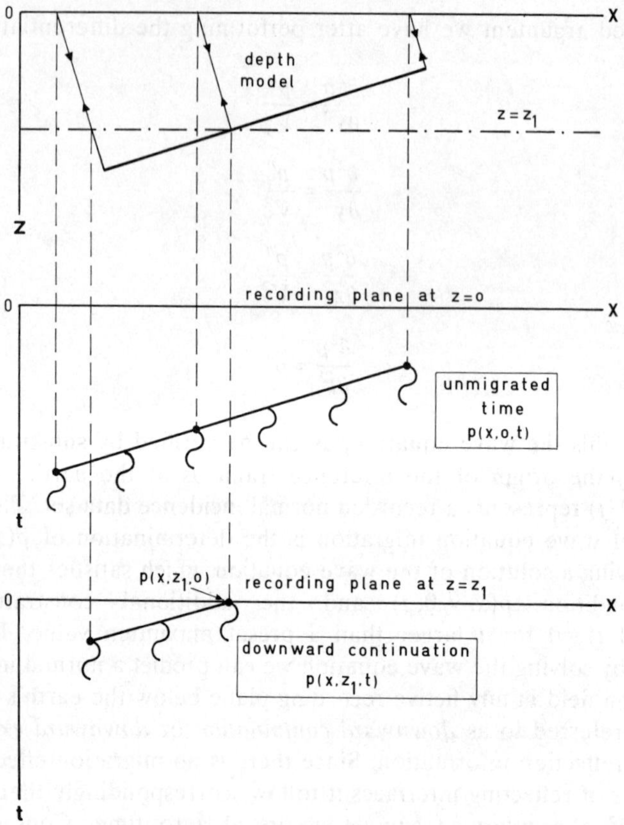

FIG. 8.4. Geometrical interpretation of downward continuation and imaging process.

The Kirchoff summation method

The Kirchoff integral for downward continuation of normal incidence reflection information from a medium of uniform velocity can be written in the form

$$p(x, y, z, t) = \frac{z}{2\pi}\int_{x'}\int_{y'}\left(\frac{1}{r^3} - \frac{2}{Vr^2}\frac{\partial}{\partial t}\right)p(x', y', 0, t+\tau)\,\mathrm{d}x'\,\mathrm{d}y'$$

The integration variables x' and y' refer to recording positions in the data acquisition plane $z = 0$, r is the distance between a fixed point $R(x, y, z)$ in the subsurface and points $S(x', y', 0)$, and τ is the two-way time $2r/V$. The expression $(\partial/\partial t)p(x', y', 0, t+\tau)$ denotes the amplitude of the differentiated time trace at a point $S(x', y', 0)$ for the time instant $t+\tau$. The variable r can be expressed in terms of the parameters x, y and z and the integration variables x' and y' by

$$r^2 = z^2 + (x - x')^2 + (y - y')^2 = z^2 + L^2$$

where L is the horizontal displacement of $S(x', y', 0)$ from $R(x, y, z)$. Accordingly we have

$$\tau^2 = T^2 + \frac{4L^2}{V^2} \qquad \text{with } T = 2z/V$$

which defines a *summation hyperboloid*.

Referring to Fig. 8.5, downward extrapolation with the Kirchoff integral proceeds as follows. In order to derive an output $p(x, y, z, t)$ for specific values of x, y, z and t, the origin of the τ–L reference frame of the summation hyperboloid is positioned at time instant t along the trace $p(x', y', 0, t)$ for which $x' = x$ and $y' = y$. Then, the quantities $p(x', y', 0, t+\tau)$ and $(\partial/\partial t)p(x', y', 0, t+\tau)$ are determined at its intersections with the set of traces $p(x', y', 0, t)$. After subsequent multiplication with the appropriate weighting factors the sum of these contributions multiplied with the factor $z/2\pi$ and trace spacings $\Delta x'$ and $\Delta y'$ yields the value of $p(x, y, z, t)$ in the extrapolated x–y–t domain. Fig. 8.5 illustrates the Kirchoff summation for time instants t_i and t_j. Keeping z fixed and repeating it for all relevant values of x, y and t will produce the complete extrapolated pressure field with migrated samples at $t = 0$. Note that during the extrapolation the shape of the summation hyperboloid is time invariant and that the origin of the time scale does not change.

Upon substitution of $z = \frac{1}{2}VT$ and $r = \frac{1}{2}V\tau$ an equivalent form of the Kirchoff integral is given by

$$\hat{p}(x, y, T, t) = \frac{2T}{\pi V^2} \int_{x'} \int_{y'} \left(\frac{1}{\tau^3} - \frac{1}{\tau^2}\frac{\partial}{\partial t}\right) p(x', y', 0, t+\tau)\, dx'\, dy'$$

Deleting the factor $1/\tau^3$ for $r \gg$ wavelength leads to the Rayleigh–Sommerfeld approximation

$$\hat{p}(x, y, T, t) = -\frac{2T}{\pi V^2} \int_{x'} \int_{y'} \frac{1}{\tau^2}\frac{\partial}{\partial t} p(x', y', 0, t+\tau)\, dx'\, dy'$$

The above can be extended to the more general case of a horizontal

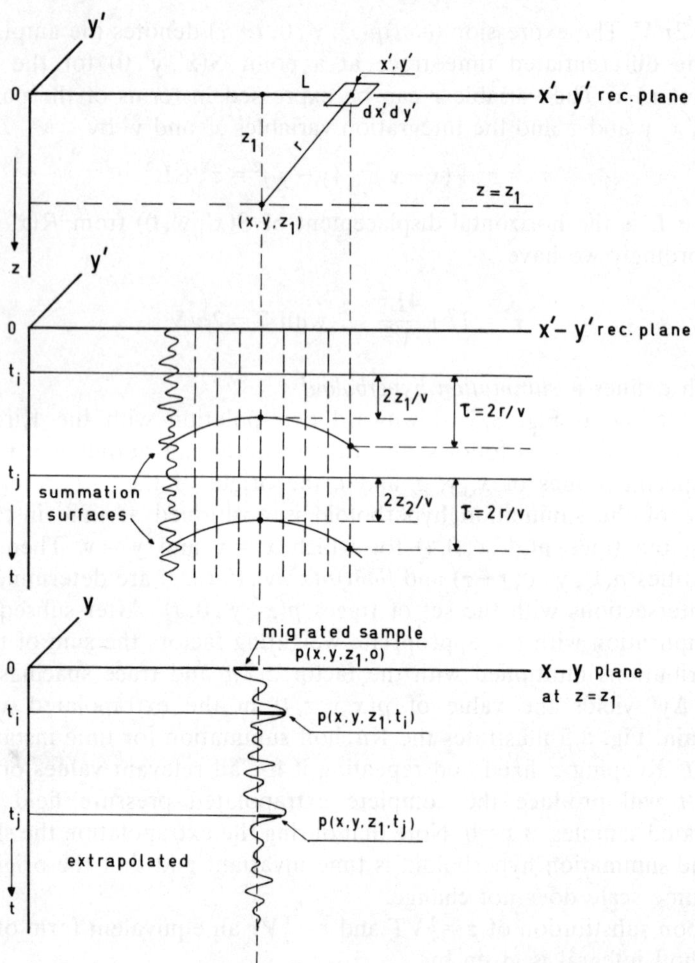

FIG. 8.5. Downward continuation and imaging based on the Kirchoff integral expression.

velocity distribution by performing the downward continuation in an iterative way. Then, considering layer thicknesses $(\Delta z)_1 \ldots (\Delta z)_p$ and velocities $V_1 \ldots V_p$, an extrapolated field at a depth $z_n = \sum_n (\Delta z)_i$ will serve as input for downward extrapolation to the depth $z_n + (\Delta z)_{n+1}$. This will yield migrated samples $p(x, y, z_n, 0)$ at times $T_n = 2 \sum_n (\Delta z)_i / V_i$.

The proper handling of non-horizontal velocity distributions is one of the main problems in post-stack signal migration. The iterative modes of wave equation migration procedures allow limited lateral velocity variations in the individual velocity layers, the finite difference method being more flexible in dealing with relatively large velocity changes. Conjectured velocity distributions should be estimated in their migrated positions, however. Uncertainties in that respect increase with structural complexity and will eventually impose a limit to the fidelity of post-stack signal migration systems.

A non-iterative Kirchoff migration in horizontal velocity distributions is based on r.m.s. velocities. The summation hyperboloid can then be approximated by

$$\tau^2 = T^2 + \frac{4L^2}{V_{\mathrm{rms}}^2(T)} \qquad \text{with } T = 2z/V_{\mathrm{av}}$$

During the non-iterative approach we shall only be interested in migrated samples $p(x, y, z, 0)$; hence, the origin of the τ–L reference frame will for all values of z be positioned at zero time. This modifies the Rayleigh–Sommerfeld approximation into

$$\hat{p}(x, y, T) = -\frac{2T}{\pi V_{\mathrm{rms}}^2(T)} \int_{x'} \int_{y'} \frac{1}{\tau^2} \frac{\partial}{\partial t} p(x', y', 0, \tau)\, \mathrm{d}x'\, \mathrm{d}y'$$

By omitting the differentiation and neglecting all weighting factors this expression reduces to

$$\hat{p}(x, y, T) = C \int_{x'} \int_{y'} p(x', y', 0, \tau)\, \mathrm{d}x'\, \mathrm{d}y'$$

which corresponds to the formulation of the unweighted diffraction stack for areal data from Section 8.1.

Two-dimensional migration

When the X'-axis extends into the direction of subsurface dip, we have for a dip profile

$$p(x', y', 0, t) = p(x', 0, 0\ t)$$

Two-dimensional migration can then be formulated by

$$\hat{p}(x, T) = -\frac{2T}{\pi V_{\mathrm{rms}}^2(T)} \int_{x'} \int_{y'} \frac{1}{\tau^2} \frac{\partial}{\partial t} p(x', 0, 0, \tau)\, \mathrm{d}x'\, \mathrm{d}y'$$

whereby $L^2 = (x - x')^2 + y'^2$.

Migration of plane wave reflection data

For plane wave reflection geometry (see Section 3.1) diffraction times are given by

$$\tau = z/V + r/V$$

$$\text{or} \qquad \tau = T/2 + [(T/2)^2 + L^2/V^2]^{\frac{1}{2}}$$

which can be approximated by

$$\tau^2 = T^2 + \frac{4L^2}{(V\sqrt{2})^2} \qquad \text{for } z \gg L$$

This suggests that for the migration of plane wave reflection profiles we may use the Kirchoff integral for zero-offset data after the velocity transformation $V' = V\sqrt{2}$.

Downward extrapolation in terms of convolution

From the definition of the delta function we have in general for a signal $S(t)$

$$S(t) = \int_{t'} \delta(t - t')S(t')\,\mathrm{d}t'$$

so that

$$p(x', y', 0, t + \tau) = \int_{t'} p(x', y', 0, t')\delta(t - t' + \tau)\,\mathrm{d}t'$$

Upon substitution of this expression into the Kirchoff integral it obtains that

$$p(x, y, z, t) = \frac{z}{2\pi}\int_{x'}\int_{y'}\int_{t'}\left(\frac{1}{r^3} - \frac{2}{Vr^2}\frac{\partial}{\partial t'}\right)$$
$$\times\, p(x', y', 0, t')\delta(t - t' + \tau)\,\mathrm{d}x'\,\mathrm{d}y'\,\mathrm{d}t'$$

which is a three-dimensional version of the convolution integral of Section 5.2, since both r and τ are functions of $(x - x')$ and $(y - y')$.

Example 8.1 *Non-iterative Kirchoff migration*

The normal incidence reflection response of a horizontal interface in the velocity distribution $V_z = 1500 + 0{\cdot}5z$ m/s is given by $p(t) = \sin^2(50\pi t)$ for $2{\cdot}000 < t < 2{\cdot}020$ s, and $p(t) = 0$ outside this time interval. Determine migrated signal amplitudes for $T = 2{\cdot}000$, $2{\cdot}005$,

2·010, 2·015 and 2·020 s from the Rayleigh–Sommerfeld approximation of the Kirchoff integral, using r.m.s. velocities. Trace spacing: $\Delta x = \Delta y = 25$ m.

Example 8.1 *Solution*
The solution of this example was obtained with a programmable calculator. The first step is the computation of V_{rms} values for the given migrated times, using the material of Sections 3.4 and 3.8; these are tabulated as follows;

T	V_{rms} (m/s)
2·000	1 966·2
2·005	1 967·7
2·010	1 969·1
2·015	1 970·6
2·020	1 972·0

Differentiation of the unmigrated signal yields $\mathrm{d}p/\mathrm{d}t = 50\pi \sin(100\pi t)$. The summation process is summarized as follows:

Time T	$\sum (1/\tau^2) 50\pi \sin(100\pi\tau)$	$\times \frac{-2T}{\pi V_{rms}^2} \times 25^2$ = *output amplitude*	$p(t) = \sin^2(50\,\pi t)$ = *input amplitude*
2·000	52	−0·011	0·0
2·005	−2 443	0·503	0·5
2·010	−4 833	0·997	1·0
2·015	−2 430	0·502	0·5
2·020	0	0·000	0·0

The finite difference method
The expression 'finite difference method' refers to a procedure for solving the wave equation with the aid of a numerical technique applied to sampled reflection information. It can be shown that the difference version of the wave equation cannot be solved without information on $\partial p/\partial z$, which is an unobserved quantity during seismic data acquisition. An approximate solution which satisfies the recorded pressure field $p(x, y, 0, t)$ may be obtained, however, after a *modification* and subsequent *simplification* of the original wave equation.

The modification of the wave equation is the heart of the finite difference method and is based on the concept of a floating time

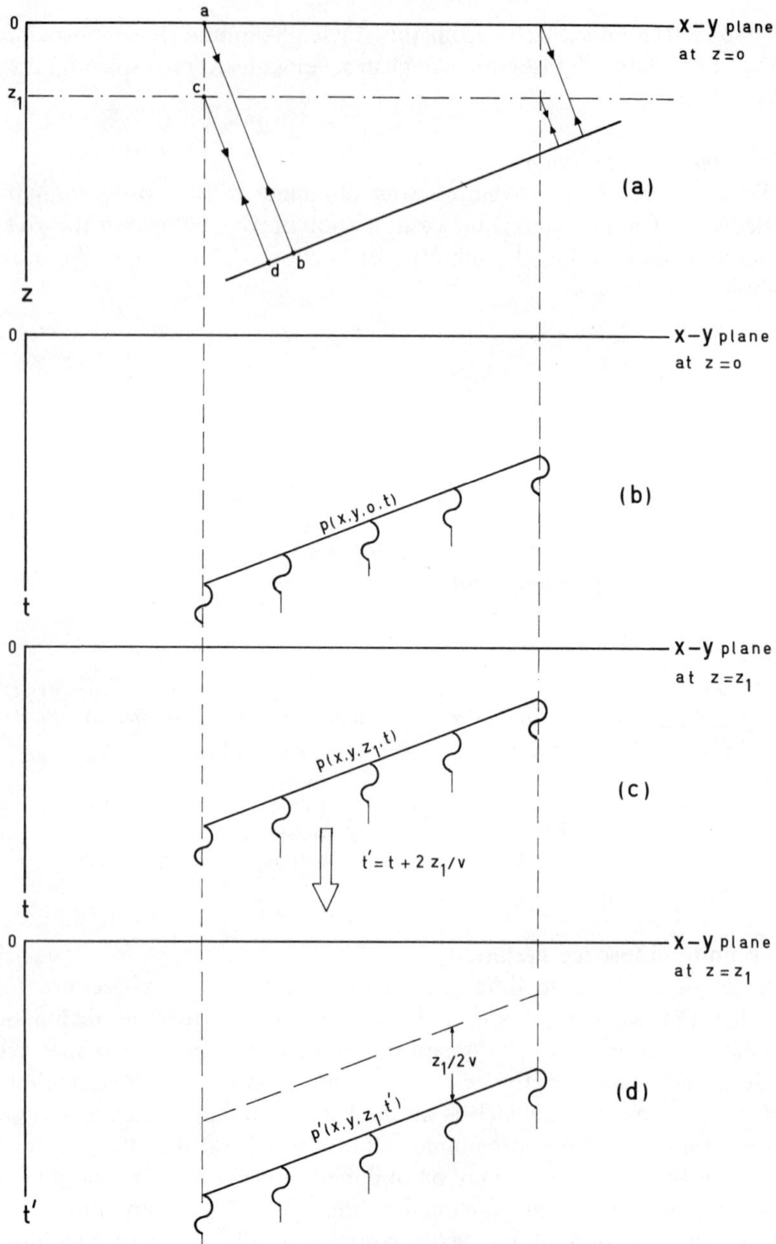

FIG. 8.6. Geometrical interpretation of floating time reference.

reference. This involves the transformation

$$t' = t + \frac{2z}{V}$$

which converts a pressure field $p(x, y, z, t)$ into $p'(x, y, z, t')$, migrated samples being represented by $p'(x, y, z, 2z/V)$. Fig. 8.6 is a geometrical interpretation of the floating time reference. Fig. 8.6(b) shows the normal incidence reflection response of the model in Fig. 8.6(a), in the plane $z=0$. We then have $t'=t$ and $p(x, y, 0, t) = p'(x, y, 0, t')$. Fig. 8.6(c) is the reflection response $p(x, y, z_1, t)$ in a fictive recording plane at the depth z_1. Finally, Fig. 8.6(d) represents the pressure field $p'(x, y, z_1, t')$ after the transformation $t' = t + 2z/V$. From the geometry, $t_{ab} < t_{acd}$; hence, $p'(x, y, 0, t')$ is not identical to $p'(x, y, z_1, t')$. Their difference is small, however, for gentle to moderate dips and becomes negligible when the dip angle tends to zero. Consequently, it obtains that

$$p'(x, y, z_i, t') \approx p'(x, y, z_j, t')$$

so that, with the above restriction

$$\frac{\partial^2 p'}{\partial z^2} \approx 0$$

The derivation of the modified and simplified wave equations proceeds now as follows. Since the pressure amplitude does not change after the transformation of the time variable we have

$$p(x, y, z, t) = p'(x, y, z, t')$$

from which

$$\frac{\partial^2 p}{\partial x^2} = \frac{\partial^2 p'}{\partial x^2} \quad \text{and} \quad \frac{\partial^2 p}{\partial y^2} = \frac{\partial^2 p'}{\partial y^2}$$

Further, in view of $t' = t + 2z/V$ it follows from the chain rule for differentiation that

$$\frac{\partial^2 p}{\partial t^2} = \frac{\partial^2 p'}{\partial t'^2}$$

and

$$\frac{\partial^2 p}{\partial z^2} = \frac{\partial^2 p'}{\partial z^2} + \frac{4}{V}\frac{\partial^2 p'}{\partial z\, \partial t'} + \frac{4}{V^2}\frac{\partial^2 p'}{\partial t'^2}$$

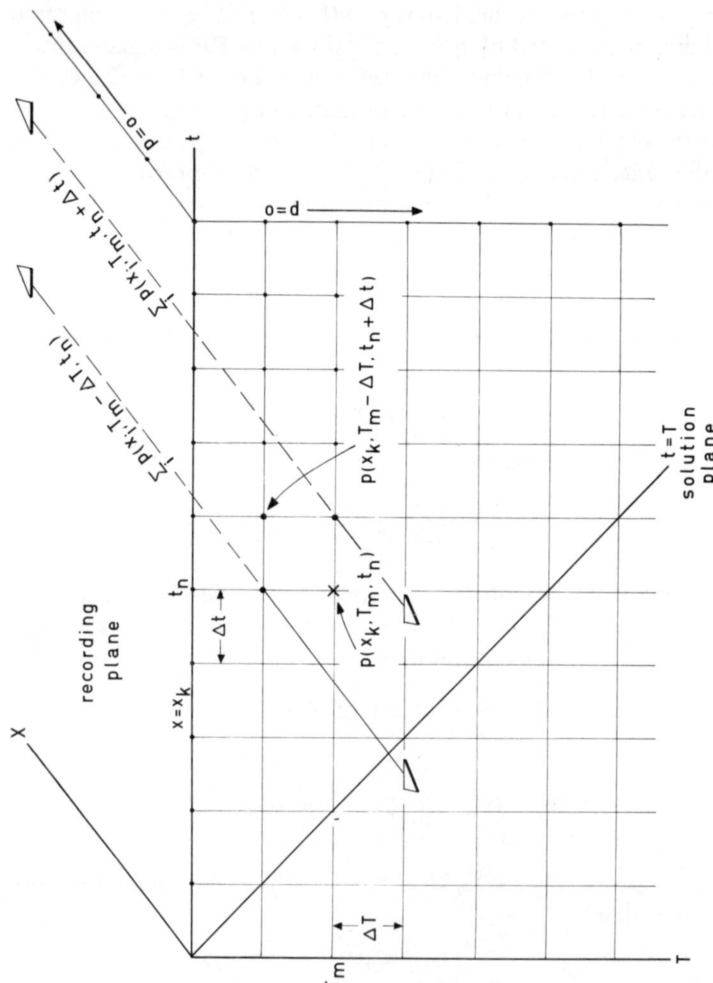

FIG. 8.7. Illustration of finite differences scheme during solution of the wave equation.

Substitution of the above expressions into the original wave equation for two-way travel times leads to the modified wave equation

$$\frac{\partial^2 p'}{\partial x^2}+\frac{\partial^2 p'}{\partial y^2}+\frac{\partial^2 p'}{\partial z^2}+\frac{4}{V}\frac{\partial^2 p'}{\partial z\,\partial t'}=0$$

which upon substitution of $\partial^2 p'/\partial z^2=0$ simplifies to

$$\frac{\partial^2 p'}{\partial x^2}+\frac{\partial^2 p'}{\partial y^2}+\frac{4}{V}\frac{\partial^2 p'}{\partial z\,\partial t'}=0$$

for dips to about 15°. Expressing the depth z in terms of migrated time T by $z=\frac{1}{2}VT$, we have

$$\frac{\partial p'}{\partial z}=\frac{2}{V}\frac{\partial p'}{\partial T}$$

so that an equivalent form of the above simplified wave equation is given by

$$\frac{\partial^2 p}{\partial x^2}+\frac{\partial^2 p}{\partial y^2}+\frac{8}{V^2}\frac{\partial^2 p}{\partial T\,\partial t}=0$$

replacing for convenience of writing the previously used symbols p' and t' by p and t.

The X–T–t block diagram in Fig. 8.7 illustrates the general procedure of solving the simplified wave equation by means of a recursive scheme for the two-dimensional case ($\partial^2 p/\partial y^2=0$). Through a digital, velocity dependent filtering process in the direction of the X-axis a sample $p(x_k, T_m, t_n)$ is derived from the sample $p(x_k, T_m-\Delta T, t+\Delta t)$ and the sample sequences $\sum_i p(x_i, T_m-\Delta T, t_n)$ and $\sum_i p(x_i, T_m, t_n+\Delta t)$. This implies that the recursion must start in the X–T plane for which $p(t, T, x)=0$, in accordance with the earlier mentioned condition that $p(x, y, 0, t)=0$ for $t>t_{max}$. The consecutive computation steps produce downward extrapolated reflection sections in X–t planes corresponding to specific values of T (or equivalent depth z), migrated samples being situated in the solution plane $t=T$.

8.3 PRE-STACK MIGRATION

Pre-stack migration is the direct transformation of multi-coverage reflection field data into migrated time or depth configurations. Its merits are twofold. First, stacking response is independent of reflector attitude since no stacking velocities are involved. Second, resolution is enhanced where owing to non-hyperbolic moveouts the conventional stack is distorted or lost in the noise background. Pre-stack migration may therefore be used as an interpretational aid in defining structural details underneath complex velocity overburdens. Early programs were designed for migration in horizontal velocity distributions, employing r.m.s. velocities. Current systems are generally of an iterative nature and can effectively handle non-horizontal velocity layering. Their application is restricted, however, to migration along dip profiles. An important attribute of pre-stack migration is the inherent possibility of deriving interval velocity information from intermediate migration results.

Generalized migration geometry in horizontal velocity layers

Fig. 8.8 illustrates a graphical migration procedure for non-zero offset c.m.p. data. A reflection point R is located on a surface of equal reflection time defined by the source–detector offset x, the reflection time t and the velocity distribution $V(z)$. This is also referred to as an aplanatic surface, which would be an ellipsoid in a medium of uniform velocity, source and detector being its focal points. When we have n-fold coverage there will be a set of n of these surfaces for each c.m.p. along a grid of reflection traverses, migrated reflector elements aligning along their common envelope, in analogy with the migration of normal incidence reflection data. The bottom of an aplanatic surface represents an unmigrated reflection point R′. Unmigrated time τ is defined as the two-way vertical time to R′. For two-dimensional data we have in general $t=f_t(x, u)$ and $\tau=f_\tau(x, t)$, where u is the position coordinate of the c.m.p. For a fixed value of x the unmigrated time gradient $\partial\tau/\partial u$ can be written in the form

$$\frac{\partial\tau}{\partial u}=\frac{\partial\tau}{\partial t}\frac{\partial t}{\partial u}$$

The derivative $\partial\tau/\partial t$ is equal to $1/\cos i_z$, where i_z is the direction of the ray SR′ at the point R′. Hence, from Section 3.6

$$\frac{\partial\tau}{\partial u}=\frac{1}{\cos i_z}\left[\frac{\sin\theta_s}{V_0}+\frac{\sin\theta_r}{V_0}\right]$$

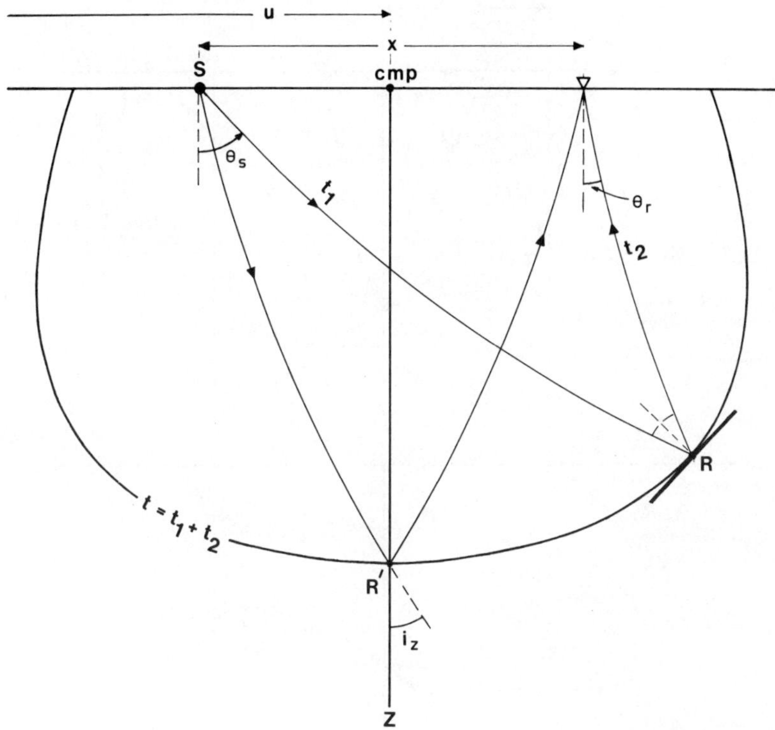

FIG. 8.8. Curve of equal reflection time (aplanatic curve) in horizontal velocity distribution.

With reference to Fig. 8.9 a diffraction curve can be expressed by

$$t = t_1 + t_2 = \left[\frac{\tau_0^2}{4} + \frac{(L + \frac{1}{2}x)^2}{V_{\text{rms}}^2(\tau_0)}\right]^{\frac{1}{2}} + \left[\frac{\tau_0^2}{4} + \frac{(L - \frac{1}{2}x)^2}{V_{\text{rms}}^2(\tau_0)}\right]^{\frac{1}{2}} = \left[\tau_L^2 + \frac{x^2}{V_{\text{rms}}^2(\tau_L)}\right]^{\frac{1}{2}}$$

Note that t in the above relation is the actual diffraction time and τ_L the unmigrated diffraction time plotted at the distance L from the central axis of the diffraction curve. For $x = 0$ we obtain again the equation of the m.a.c. curve from Section 8.1.

Placing the diffraction source at the reflection point R in Fig. 8.8, unmigrated reflection and diffraction time gradients will be identical in view of common angles of emergence of reflection and diffraction raypaths at source and detector positions. For each value of the shooting

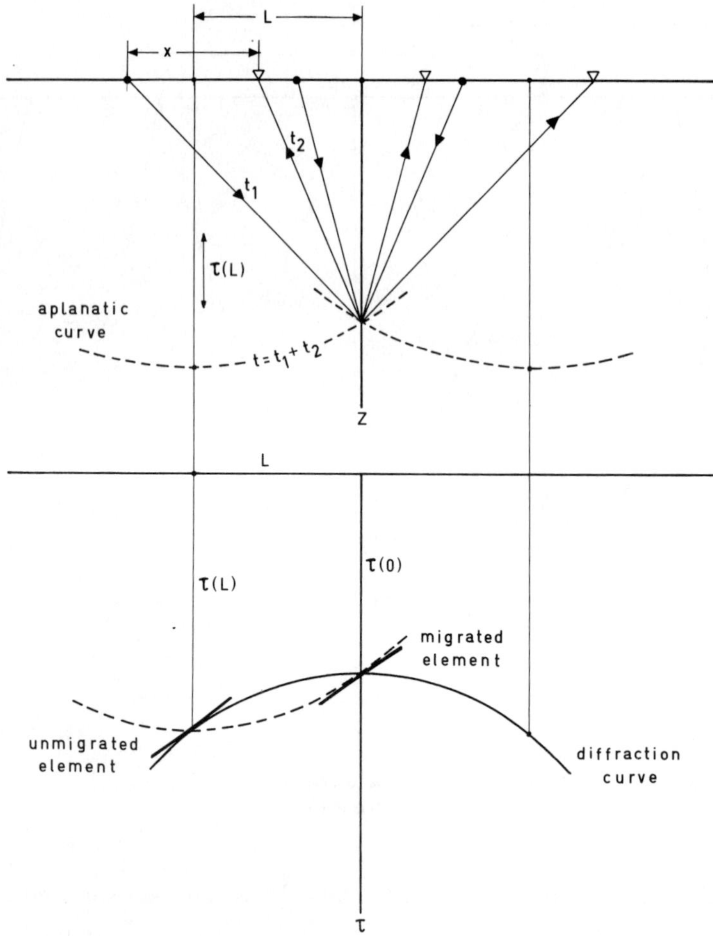

Fig. 8.9. Two-dimensional migration geometry in horizontal velocity distribution for common offset gather. Diffraction curve for source–detector spacing x.

distance x an unmigrated reflector in the τ-domain is therefore the envelope of a sequence of diffraction curves, in analogy with zero-offset migration geometry in Section 7.5. Correspondingly, digital pre-stack migration in horizontal velocity distributions can be described as the summation of unmigrated signal samples from common offset gathers along generalized m.a.c. curves. This will produce again an unweighted diffraction stack.

Example 8.2 *Diffraction curve*

1. Plot a L–τ diffraction curve for $x = 750$ m and $L_{max} = 2000$ m from a diffraction source at a depth of 1500 m in a 2000 m/s velocity medium.
2. Determine the unmigrated diffraction time gradient when L is 1000 m.
3. Derive the corresponding migrated time gradient for a reflector element at the diffraction source.

Example 8.2 *Solution*

1. The diffraction curve passes through the following points

L	τ
0	1·500
500	1·577
1 000	1·791
1 500	2·105
2 000	2·482

2. For $L = 1000$ m we have $\cos i_z = 0{\cdot}980\,44$, $\sin \theta_r = 0{\cdot}384\,61$ and $\sin \theta_s = 0{\cdot}675\,72$, from which $\partial\tau/\partial u = 0{\cdot}541$ s/km.
3. The equation of an aplanatic curve for a c.m.p. at $L = 1000$ m can be written in the form

$$4z^2 = (V^2t^2 - x^2)\left(1 - \frac{4L'^2}{V^2t^2}\right) \qquad \text{with } L' = 1000 - L$$

The diffraction time at $L = 1000$ m is 1·830 s. After differentiation and upon substitution of $t = 1{\cdot}830$, $L' = 1000$ and the relevant values for x and V we find for the structural dip of the reflector element $\tan^{-1} 0{\cdot}638\,67$. For $V = 2000$ m/s its migrated time gradient is then 0·639 s/km.

Pre-stack migration by modelling

For non-zero offset reflection data, unmigrated times and diffraction τ–L curves lose their meaning in media where the velocity varies laterally. Diffraction curves can then no longer be expressed in terms of r.m.s. velocities; they become asymmetrical and do not culminate at migrated points, in analogy with normal incidence migration geometry in Fig. 7.20. Under these conditions the migration property that is preserved, however, is that migrated reflector elements are tangent to aplanatic surfaces. Their determination in arbitrary velocity distributions is therefore the principal problem of pre-stack migration. This can be accomplished through a modelling procedure, which is illustrated in Fig. 8.10 for the two-dimensional case. An assumed velocity

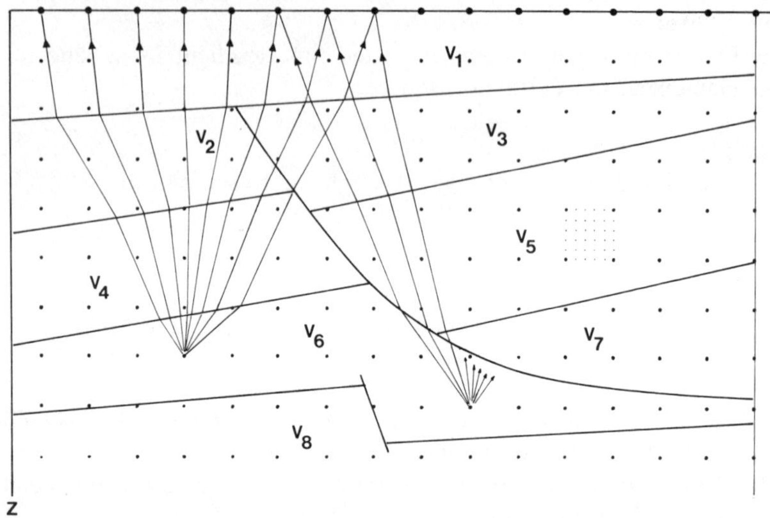

FIG. 8.10. Example of gridded velocity model for derivation of aplanatic curves in non-horizontal velocity distribution. Fermat times are computed from each grid point in the subsurface to groups of data points along the ground surface.

distribution model is gridded, after which Fermat times are computed from each of its grid points to data points along the ground surface. The grid spacing in the horizontal direction should be one detector station interval, that in the vertical direction normally being of the order of 50 m. The computation of Fermat times may be based on a minimum time algorithm similar to that discussed in Section 6.2 for the evaluation of well shoot information.

With reference to Fig. 8.11, an aplanatic curve for a reflection time t and a source detector offset x can be derived from the above Fermat times by searching along grid columns for points $P(L, z)$ so that $t_1 + t_2 = t$, t_1 and t_2 being the travel times from P to source and detector positions which straddle the relevant c.m.p. at the fixed distance x. In that manner aplanatic curves can be computed for all sample positions

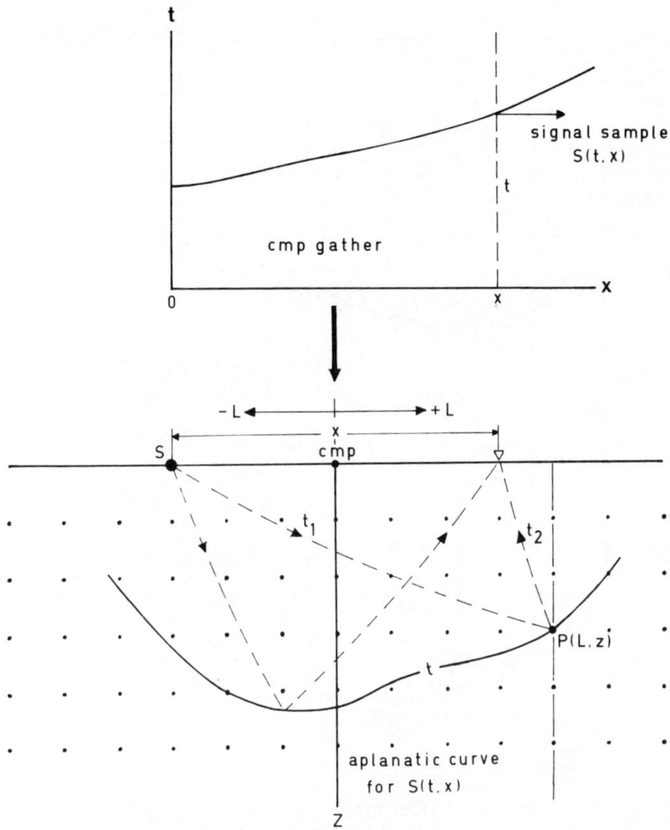

FIG. 8.11. Computation of aplanatic curve for source–detector spacing x.

along a complete set of c.m.p. traces. Two-dimensional migration is then performed by distributing signal samples along families of aplanatic curves along the reflection profile, comparable to the wavefronts produced during post-stack migration in the time domain.

A check on the validity of an assumed velocity distribution model is based on the consideration that for all sets of common x gathers the migration output should be identical. Such a consistency of migration results can be examined on displays of *migration gathers*, whose geometry is illustrated in Fig. 8.12. Point U represents a specific c.m.p. for which we shall derive a migration gather. The distances L to neighbouring c.m.p.s are measured with reference to U. Aplanatic

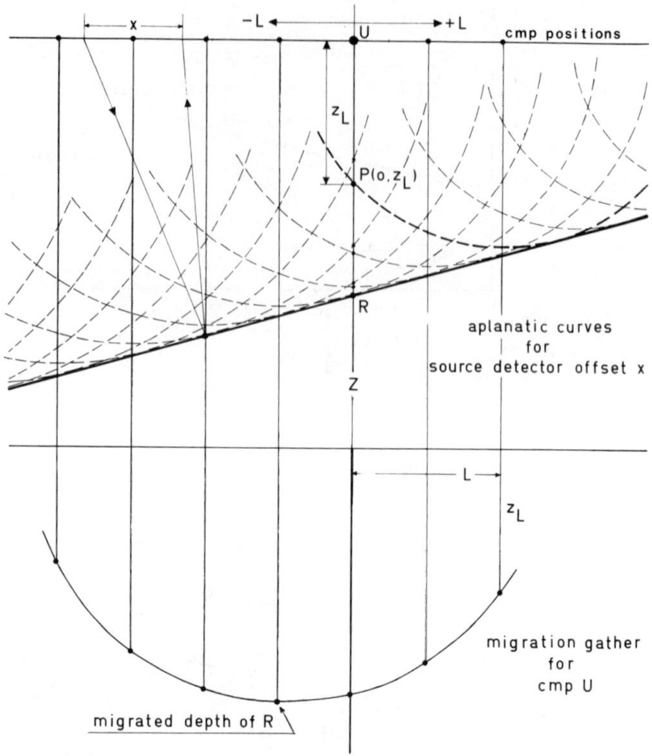

FIG. 8.12. Geometry of migration gather.

curves for a fixed source–detector offset and tangent to a single reflecting interface intersect the depth axis through U at points P(0, z_L). The curve in the lower part of Fig. 8.12 represents the migration gather in the form of an L versus z_L plot. Note that the depth to its deepest point is identical to the migrated depth of reflection point R. This implies that at point U the bottoms of migration gathers for all source–detector offsets x should line up horizontally when displayed side by side. Dip in the direction of increasing values of x signifies an overburden velocity that is too large and vice versa, providing a means for updating initial velocity estimates. Velocity diagnosis may also be based on a side by side display of stacked migration gathers, each stack being a migrated reflection trace. For the special case of a diffraction event the individual migration gathers will be straight when the velocity distribution is correct; all aplanatic curves intersect then at the diffraction source. Otherwise, convex or concave gathers of diffraction events indicate that velocities are too large or too small respectively.

For continuous control on lateral velocity variations migration gather profiles or their stacked versions should be analysed for each consecutive c.m.p. along the reflection traverse. The velocity regime of a velocity unit may then be represented by an expression of the form $V(x, z) = V(0, 0) + k_x x + k_z z$. In practice, velocity checks are usually taken at larger intervals, depending on structural complexity.

Current pre-stack migration procedures in non-horizontal velocity distributions are restricted to migration along dip profiles and carried out iteratively. An initial velocity distribution model is derived from velocity information on sonic logs and the results of a rude post-stack migration converted into depth. Next, after gridding the model and the computation of Fermat times migration gathers for the first significant velocity interface are analysed and velocities updated until a satisfactory alignment of the gathers is obtained at all velocity control points along the reflection traverse. Subsequently, this process is repeated for all the deeper velocity interfaces in the range of interest. Each migration step involves a velocity analysis for a specific sedimentary unit and a recomputation of Fermat times. The final migration result can be considered as a stack of the complete assemblage of corrected migration gathers generated during the iterations. Reflection profile S is an example of a migrated depth section obtained by means of the above described procedure.

Conclusions

Pre-stack migration can be considered as a self-contained reflection interpretation system, providing information on interval velocities as well as on the true attitude of reflecting interfaces in the depth domain. Current algorithms are based on the simplifying assumption of a two-dimensional subsurface, however. This implies that where the normal incidence stack is severely distorted, or below the noise level, an exact solution of three-dimensional reflection interpretation problems can as yet not be obtained.

Example 8.3 *Migration gathers*

Referring to the system of symbols in Fig. 8.12, derive migration gathers for velocities of 2000 m/s and 2500 m/s from the common x trace gathers tabulated below. Which velocity is correct?

$x = 500$ m		$x = 1500$ m	
L (m)	t (s)	L (m)	t (s)
−1 000	0·784	−1 000	1·049
−500	0·703	−500	0·989
−100	0·638	−100	0·944
0	0·622	0	0·934
100	0·606	100	0·923
500	0·543	500	0·883
1 000	0·468	1 000	0·839

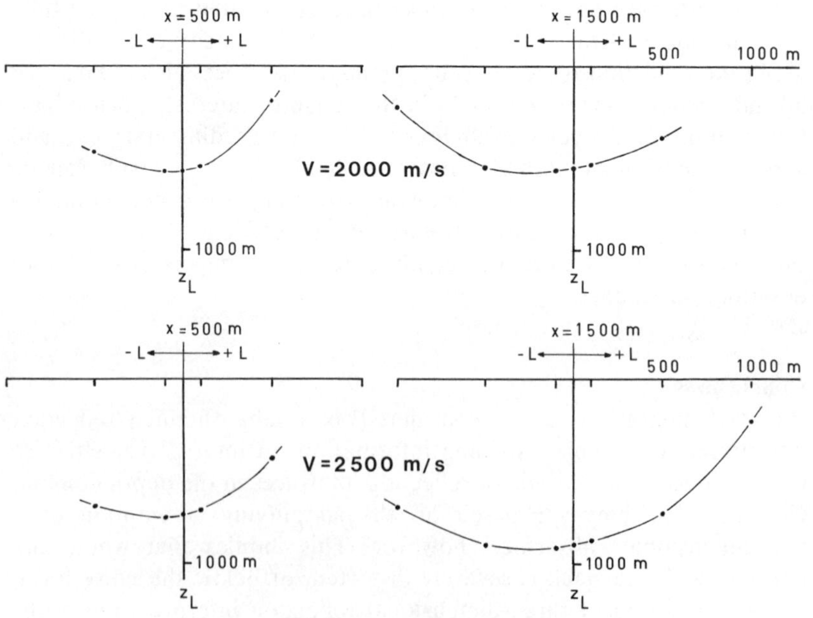

FIG. 8.13. Example 8.3 solution.

Example 8.3 *Solution*
The depths z_L are obtained from the relation

$$4z_L^2 = (V^2t^2 - x^2)\left(1 - \frac{4L^2}{V^2t^2}\right)$$

The two sets of migration gathers are shown in Fig. 8.13. The correct velocity is 2000 m/s.

BIBLIOGRAPHY

Lindsey, J. P. and Herman, A. (1970). Digital migration, *Oil & Gas Journal* (February 16).

Clearabout, J. F. (1971). Toward a unified theory of reflector mapping, *Geophysics*, **36**(3).

Clearabout, J. F. and Doherty, S. M. (1972). Downward continuation of moveout-corrected seismograms, *Geophysics*, **37**(5).

French, W. S. (1974). Two-dimensional and three-dimensional migration of model experiment reflection profiles, *Geophysics*, **39,** 265–277.

Gardner, G. H. F., French, W. S. and Matzuk, T. (1974). Elements of migration and velocity analysis, *Geophysics*, **39**(6).

Sattlegger, J. W. and Stiller, P. K. (1974). Section migration, before stack, after stack, or in-between, *Geophysical Prospecting*, **22**(2).

Alford, R. M., Kelly, K. R. and Boore, D. M. (1974). Accuracy of finite difference modelling of the acoustic wave equation, *Geophysics*, **39,** 834–842.

French, W. S. (1975). Computer migration of oblique seismic reflection profiles, *Geophysics*, **40,** 961–980.

Sattlegger, J. W., Stiller, P. K. and Echterhoff, J. A. (1976). Dip selective migration velocity determination, *Geophysical Prospecting*, **24**(4).

Loewenthal, D. L. L., Robertson, R. and Sherwood, J. W. C. (1976). The wave equation applied to migration, *Geophysical Prospecting*, **24,** 380–399.

Chander, R. (1977). On tracing seismic rays with specified end points in layers of constant velocity and plane interfaces, *Geophysical Prospecting*, **25**(1).

Hubral, P. (1977). Time migration—some ray theoretical aspects, *Geophysical Prospecting*, **25**(4).

Hood, P. (1978). Finite difference and wave number migration, *Geophysical Prospecting*, **26**(4).

Bolondi, G., Rocca, F. and Savelli, S. (1978). A frequency domain approach to two-dimensional migration, *Geophysical Prospecting*, **26**(4).

Stolt, R. H. (1978). Migration by Fourier transform, *Geophysics*, **43**(1).

Robinson, J. C. and Robbins, T. R. (1978). Dip-domain migration of two-dimensional seismic profiles, *Geophysics*, **43**(1).

Schultz, P. S. and Clearabout, J. F. (1978). Velocity estimation and downward continuation by wave front synthesis, *Geophysics*, **43,** 691–714.

Schneider, W. S. (1978). Integral formulation for migration in two or three dimensions, *Geophysics,* **43,** 49–76.

Kuhn, M. J. (1979). Acoustical imaging of source receiver coincident profiles, *Geophysical Prospecting,* **27,** 62–77.

Berkhout, A. J. and van W. Palthe, D. W. (1979). Migration in terms of spatial deconvolution, *Geophysical Prospecting,* **27**(1).

Berryhill, J. R. (1979). Wave-equation datuming, *Geophysics,* **44**(8).

Dubrulle, A. A. and Gazdag, J. (1979). Migration by phase shift—An algorithmic description for array processors, *Geophysics,* **44**(10).

Hood, P. (1979). Migration. In: *Developments in Geophysical Exploration Methods II,* Applied Science, London.

Yilmaz, O. (1979). Pre-stack partial migration, PhD Thesis, Department of Geophysics, Stanford University, Ca.

Schultz, P. S. and Sherwood, J. W. C. (1980). Depth migration before stack, *Geophysics,* **45**(3).

Judson, D. R., Lin, J., Schultz, P. S. and Sherwood, J. W. C. (1980). Depth migration after stack, *Geophysics,* **45,** 361–375.

Sattlegger, J. W., Stiller, P. K., Echterhoff, J. A. and Hentschke, M. K. (1980). Common offset plane migration (COPMIG), *Geophysical Prospecting,* **28**(6).

Berkhout, A. J. (1980). *Seismic Migration—Imaging of Acoustic Energy by Wave Field Extrapolation,* Elsevier, Amsterdam–Oxford–New York.

Clayton, R. W. and Engquist, B. (1980). Absorbing side boundary conditions for wave-equation migration, *Geophysics,* **45**(5).

Jain, S. and Wren, A. E. (1980). Migration before stack—procedure and significance, *Geophysics,* **45**(2).

Berkhout, A. J. and van W. Palthe, D. W. (1980). Migration in the presence of noise, *Geophysical Prospecting,* **28**(3).

Chun, J. H. and Jacewitz, C. A. (1981). Fundamentals of frequency domain migration, *Geophysics,* **46**(5).

Larner, K. L., Hatton, L. and Gibson, B. S. (1981). Depth migration of imaged time sections, *Geophysics,* **46**(5).

Hatton, L., Larner, K. L. and Gibson, B. S. (1981). Migration of seismic data from inhomogeneous media, *Geophysics,* **46**(5).

Berkhout, A. J. and de Jong, B. A. (1981). Recursive migration in three dimensions, *Geophysical Prospecting,* **29**(5).

Lynn, H. B. and Deregowski, S. (1981). Dip limitations on migrated sections as a function of line length and recording time, *Geophysics,* **46** (October).

Berkhout, A. J. (1981). Wave field extrapolation techniques in seismic migration, a tutorial, *Geophysics,* **46** (December).

Seismic Reflection Profiles

TIME STRATIGRAPHIC ABBREVIATIONS

T	Tertiary (clastics)
KU	Upper Cretaceous (chalk where present on profiles)
KL	Lower Cretaceous (clastics)
J	Jurassic (clastics)
TR	Triassic (clastics and evaporites)
Z	Zechstein, Upper Permian (evaporites, mainly salt, where present on profiles)
Z-3	Local dolomite deposit in Zechstein formation
R	Rotliegendes, Lower Permian (Slochteren sandstone, important gas reservoir)
C	Carboniferous
b	base of stratigraphic unit
t	top of stratigraphic unit

MAJOR UNCONFORMITIES

base Tertiary
base Lower Cretaceous (late Kimmerian unconformity)
base Rotliegendes (Saalian unconformity)

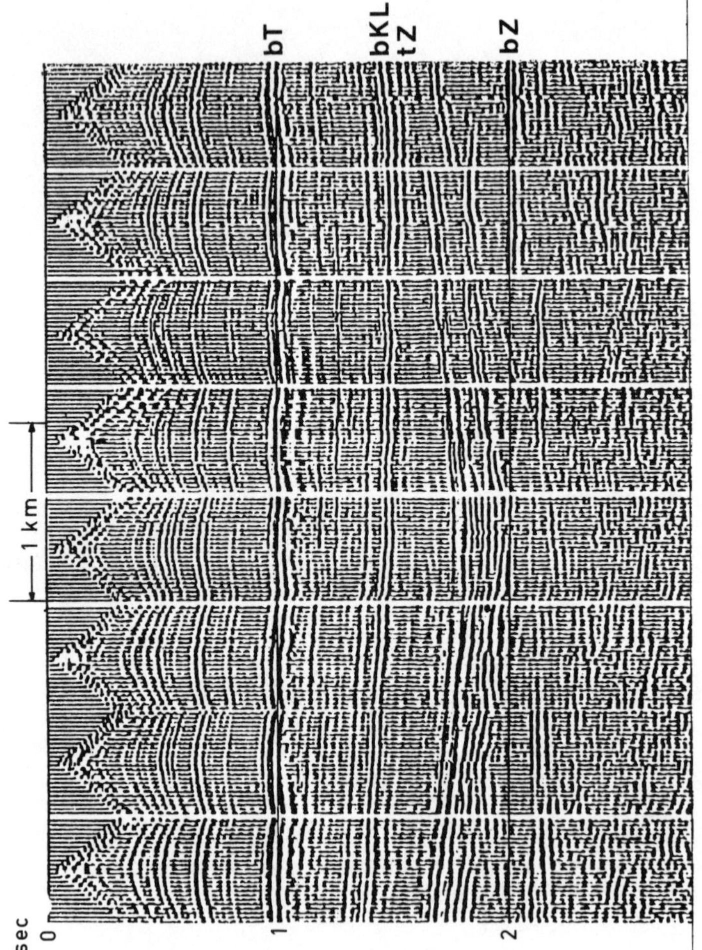

PROFILE A. Early 1960s centre-spread reflection profile, Groningen area (Netherlands).

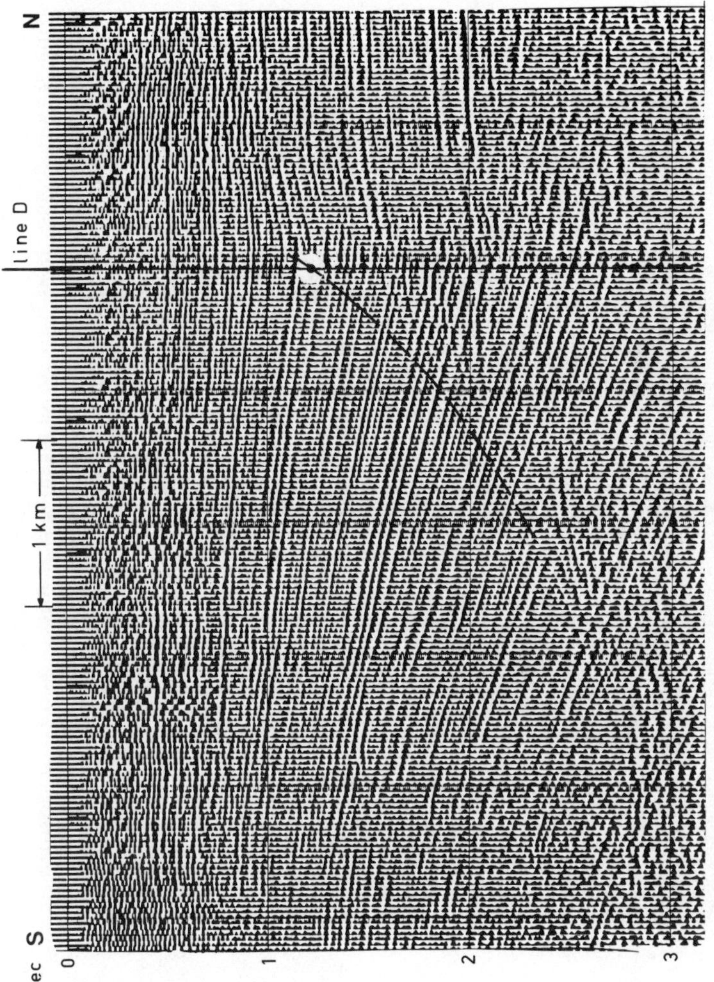

PROFILE B. Dip profile across Tertiary rollover structure bounded by major growth fault, offshore Southern Nigeria (part of data set B–C–D, Section 7.2).

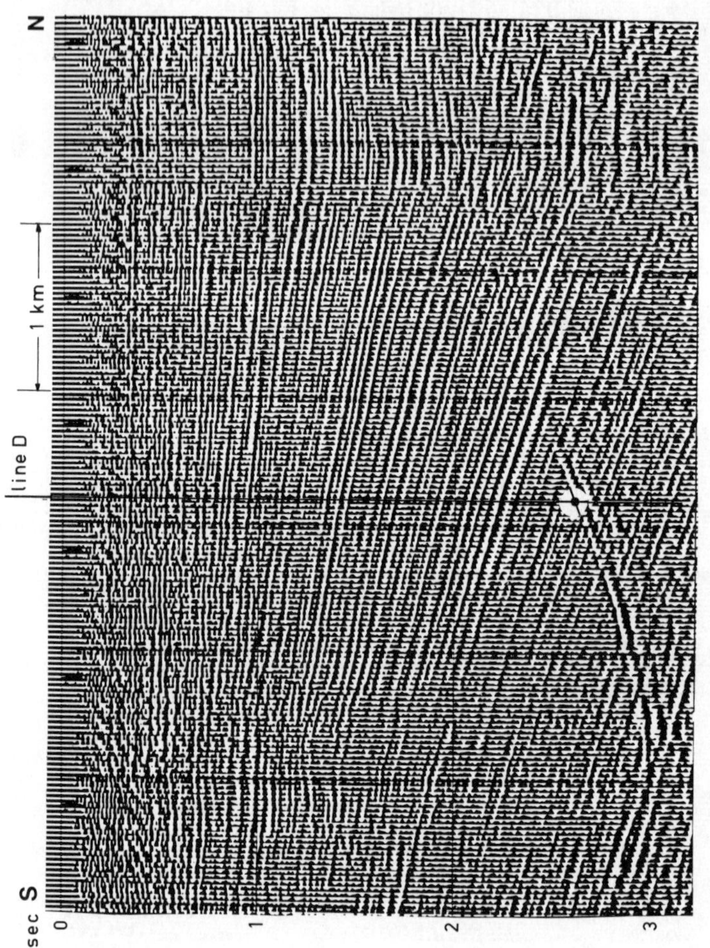

PROFILE C. Dip profile across Tertiary rollover structure bounded by major growth fault, offshore Southern Nigeria (part of data set B–C–D, Section 7.2).

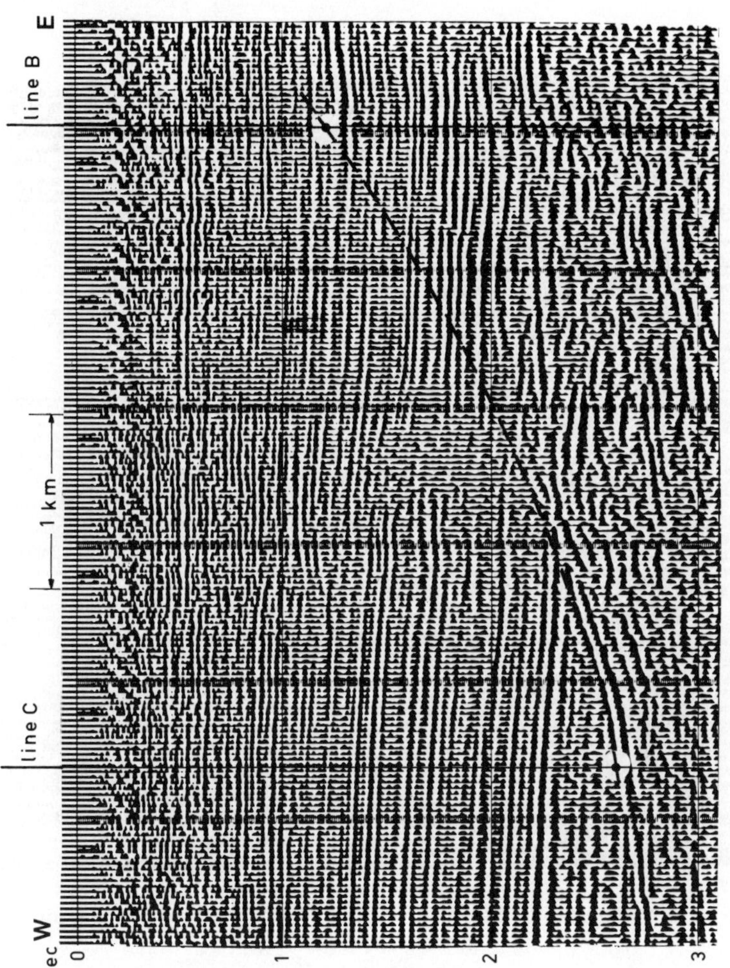

Profile D. Strike profile connecting lines B and C.

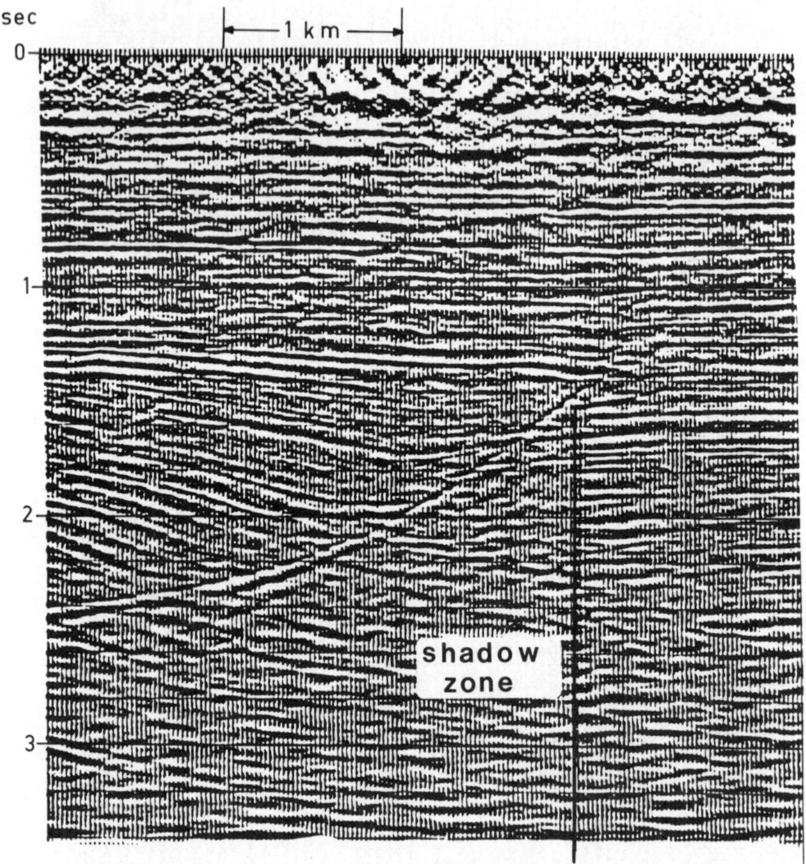

PROFILE E. Manifestation of shadow zone below major growth fault on migrated section.

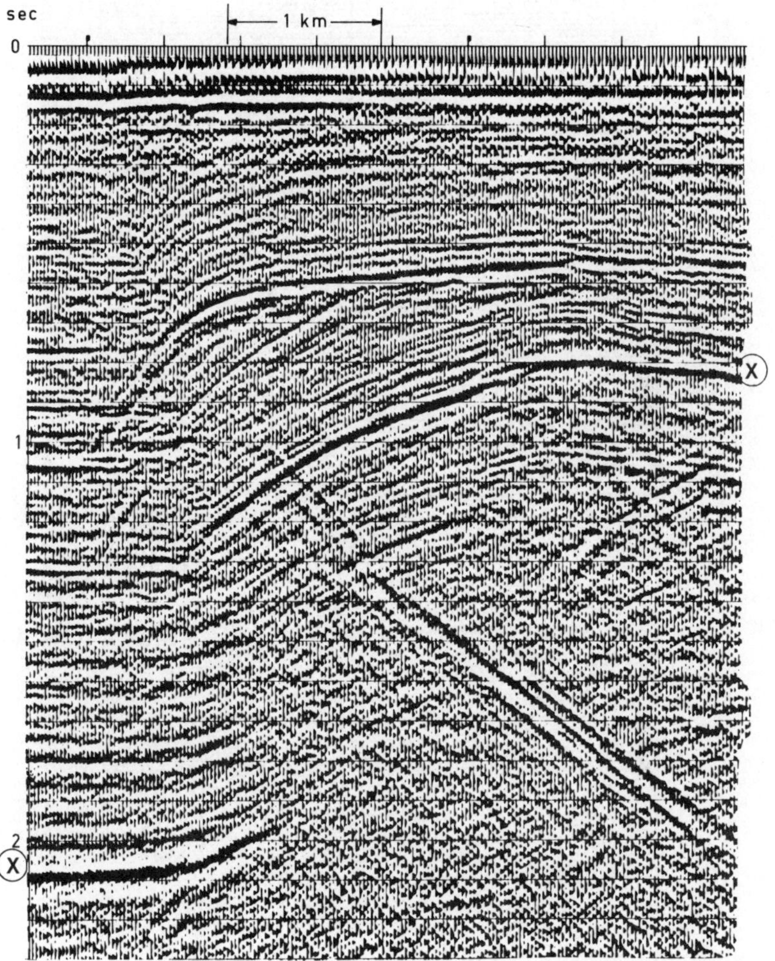

PROFILE F. Major reverse fault and shadow zone in pre-Tertiary environment, Yellow Sea (Far East).

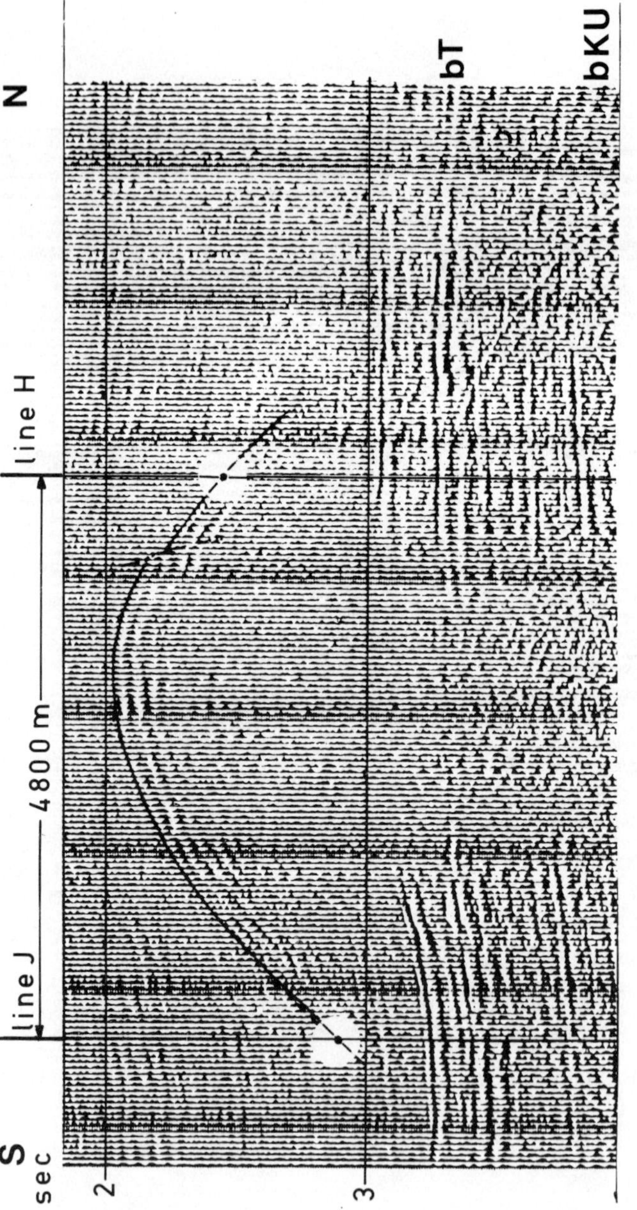

PROFILE G. Reflection profile across salt diapir, offshore Norway (part of data set G–H–I–J, Fig. 7.3, Section 7.2).

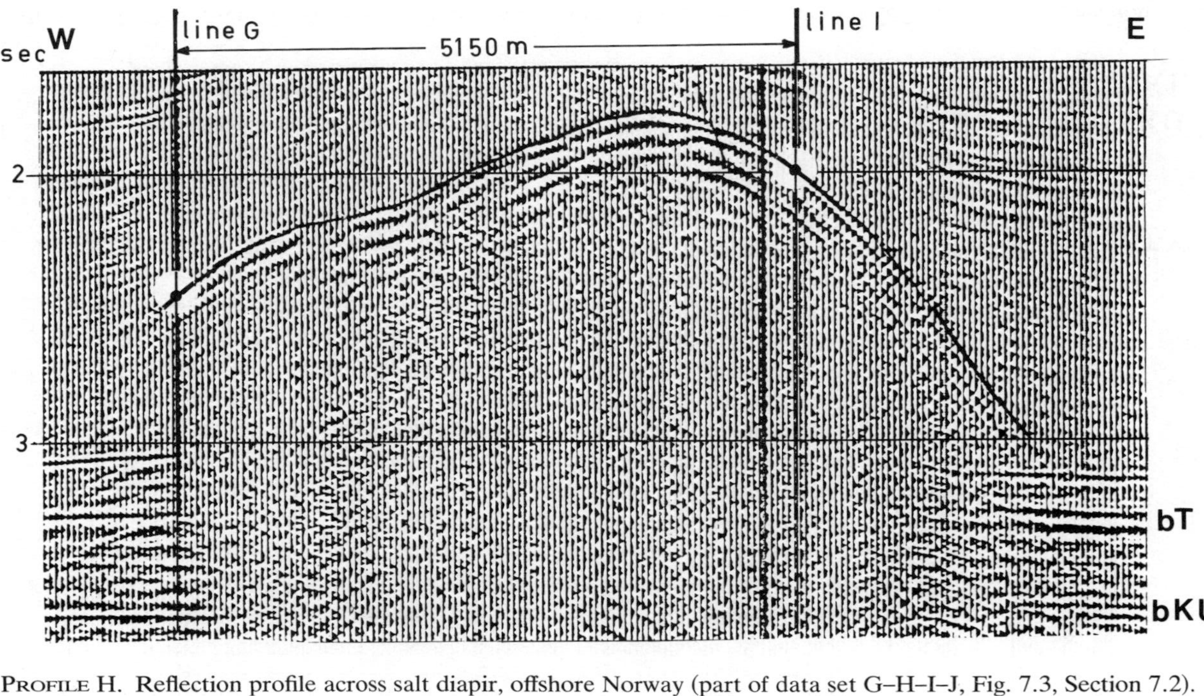

PROFILE H. Reflection profile across salt diapir, offshore Norway (part of data set G–H–I–J, Fig. 7.3, Section 7.2).

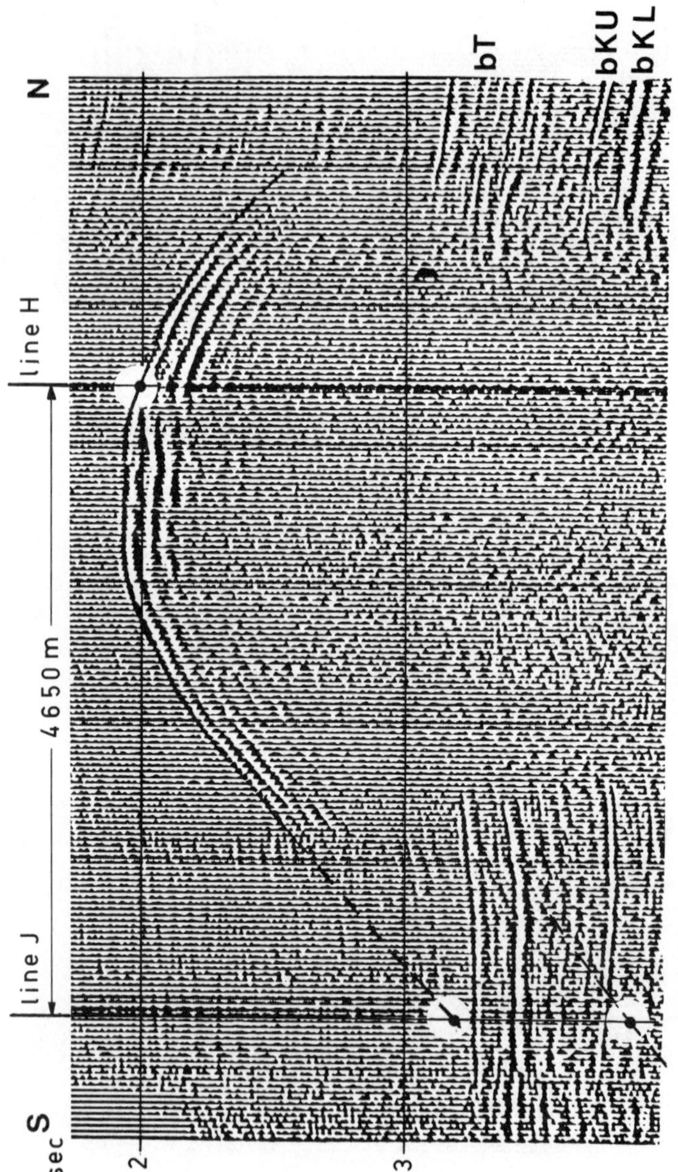

PROFILE I. Reflection profile across salt diapir, offshore Norway (part of data set G–H–I–J, Fig. 7.3, Section 7.2).

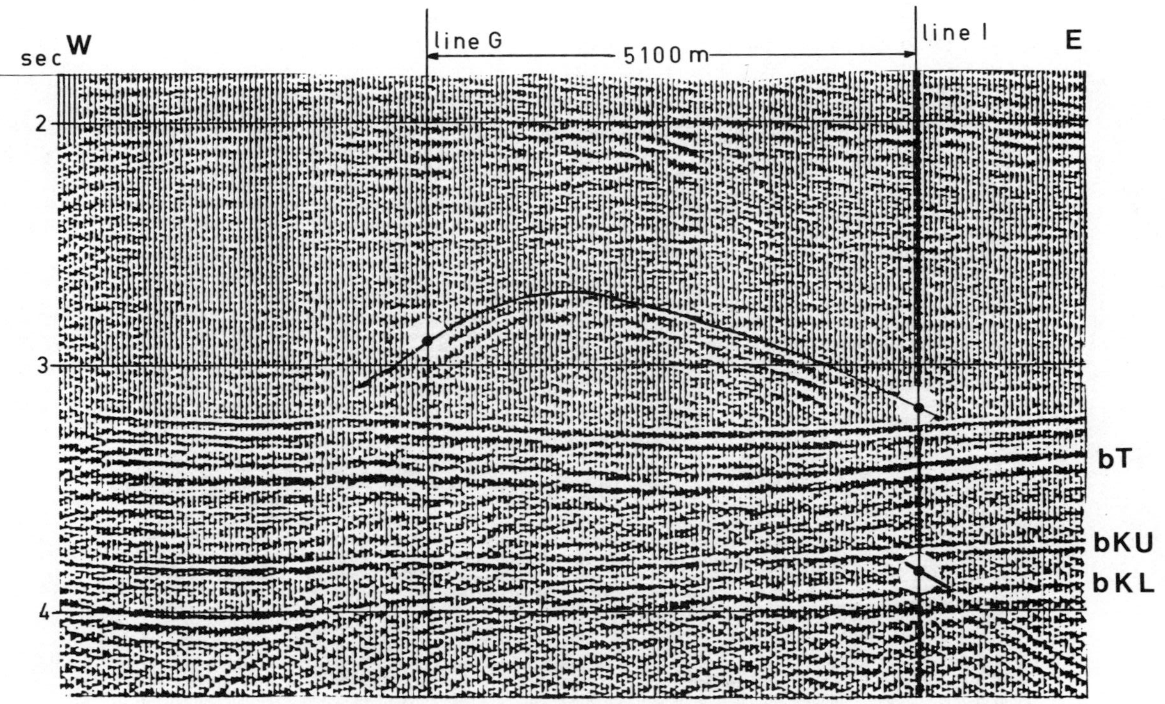

PROFILE J. Reflection profile passing near salt diapir showing cap rock reflection in side swipe position, offshore Norway (part of data set G–H–I–J. Fig. 7.3, Section 7.2).

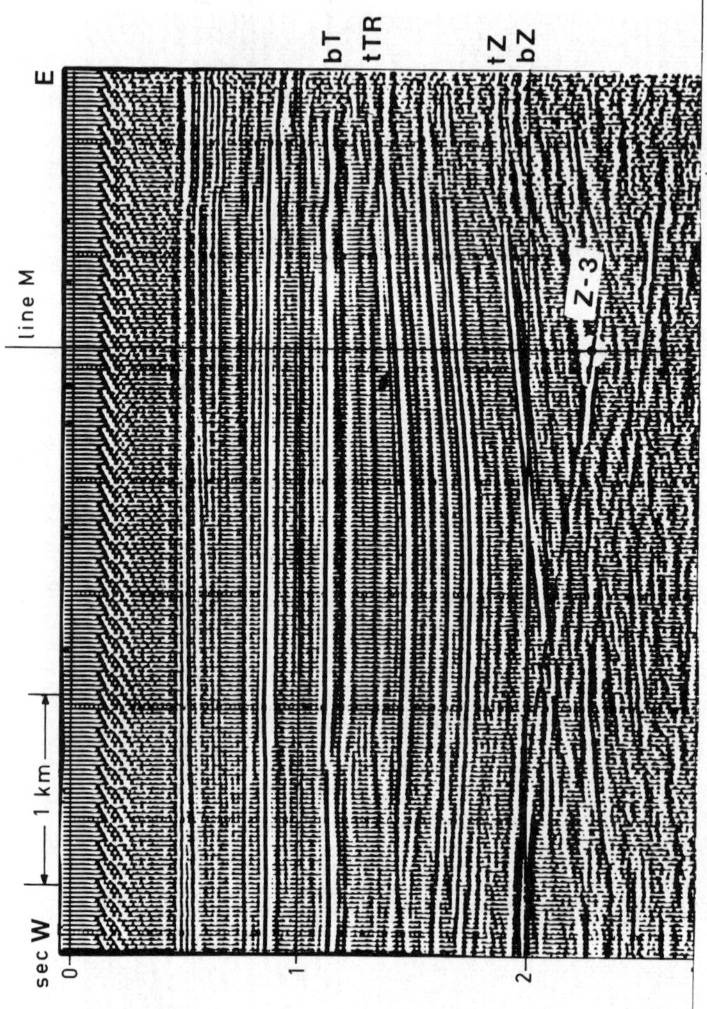

PROFILE K. Reflection profile across Z-3 anticline, offshore Netherlands (part of data set K–L–M, Section 7.2).

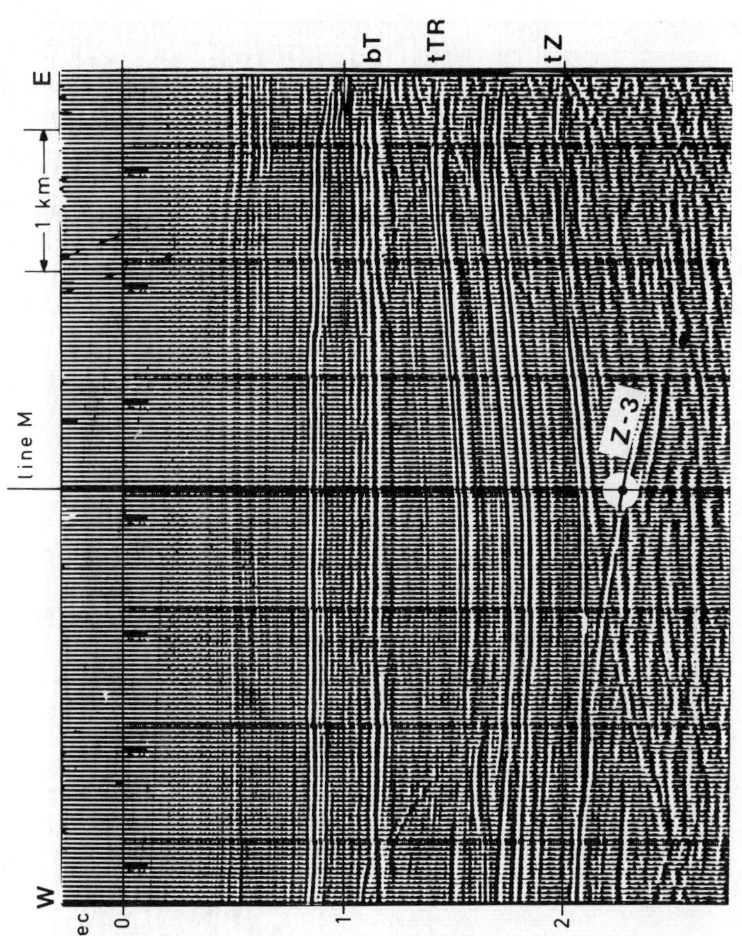

PROFILE L. Reflection profile across Z-3 anticline, offshore Netherlands (part of data set K–L–M, Section 7.2).

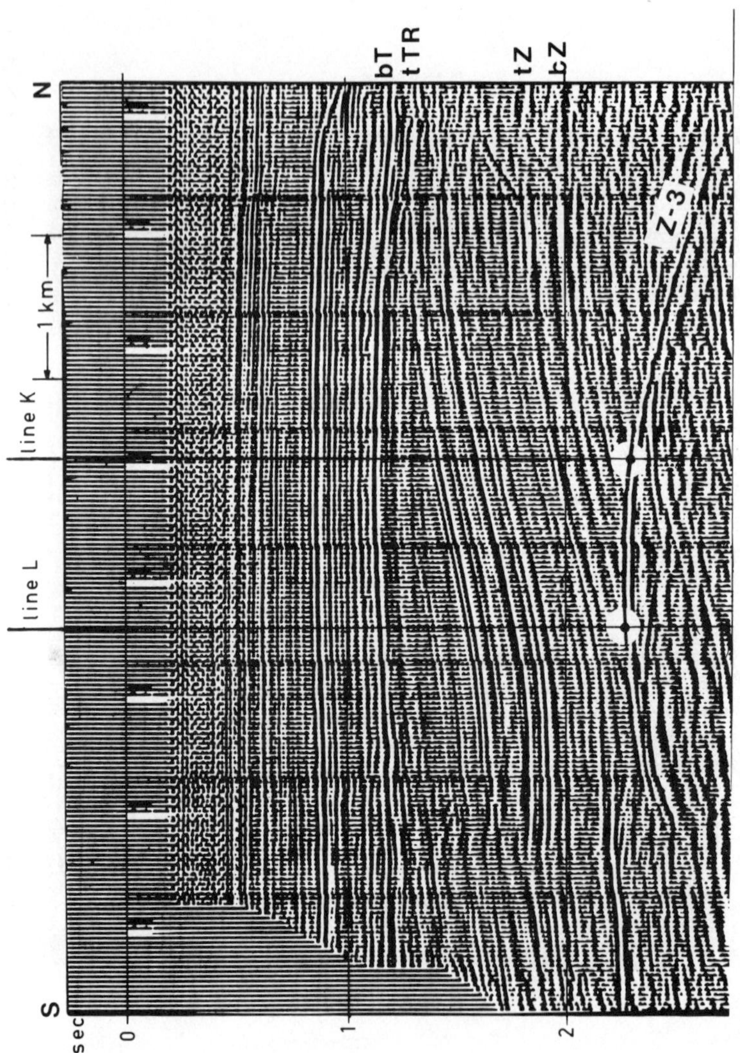

PROFILE M. Reflection profile connecting lines K and L, showing Z-3 reflection in side swipe position.

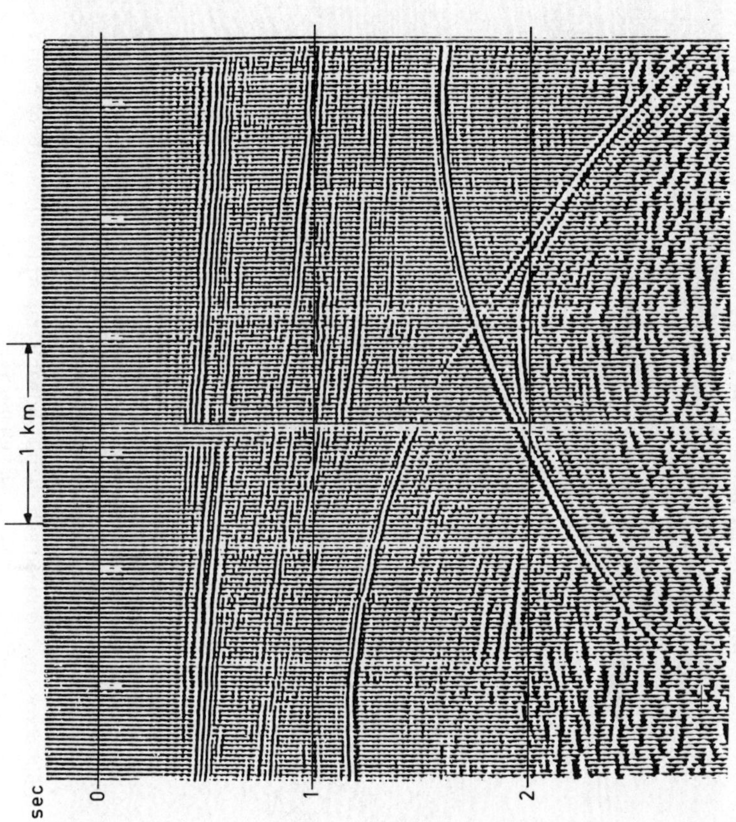

PROFILE N. Example of reflection response of pronounced synclinal feature, Adriatic Sea.

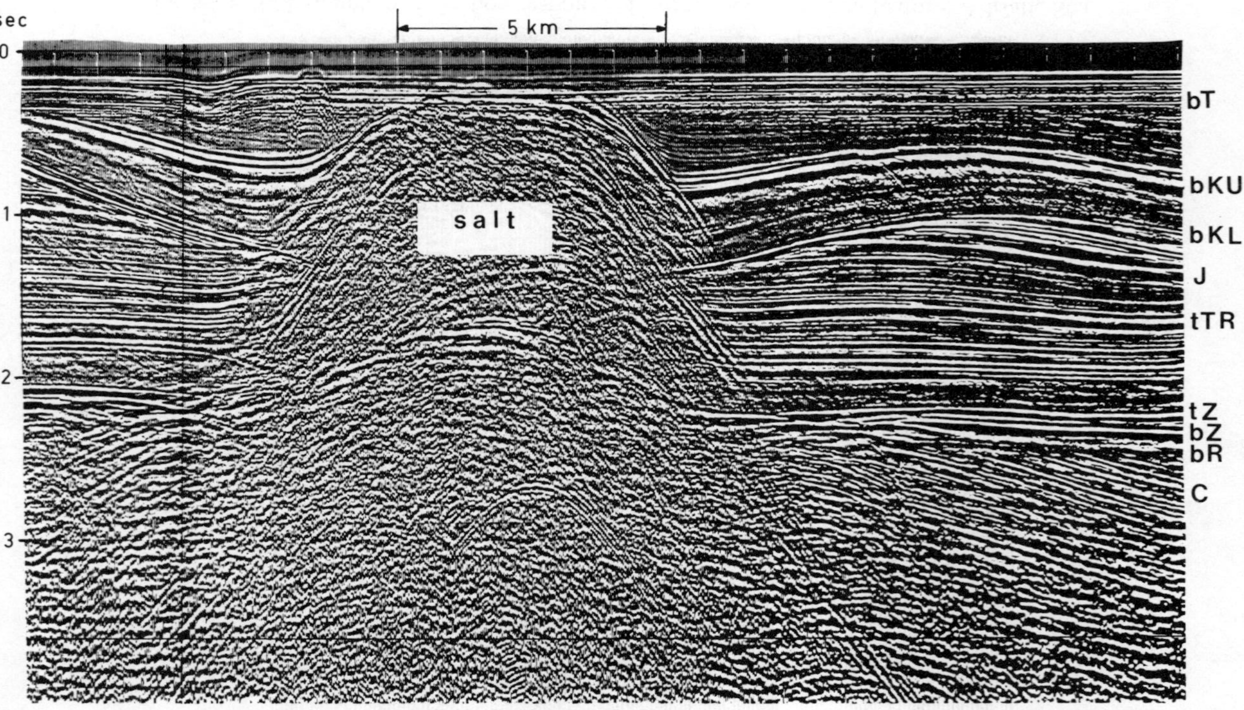

PROFILE O. Reflection profile across pronounced salt diapir showing important elements of North Sea geology, offshore England.

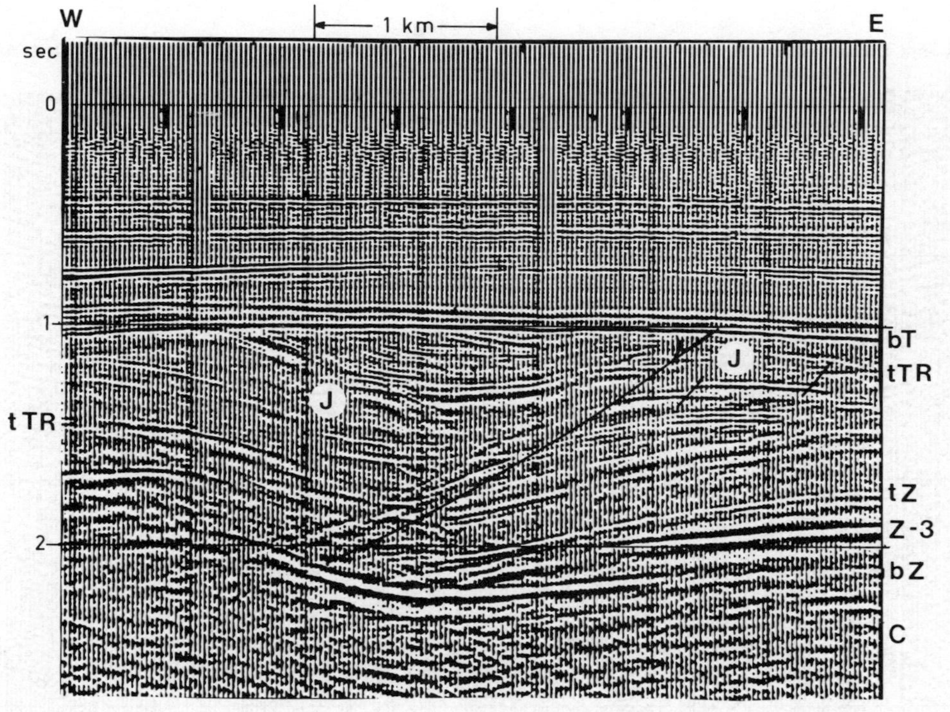

PROFILE P. Reflection profile across major Triassic fault, offshore Netherlands. Example of complicated depth conversion problem (see Section 7.6).

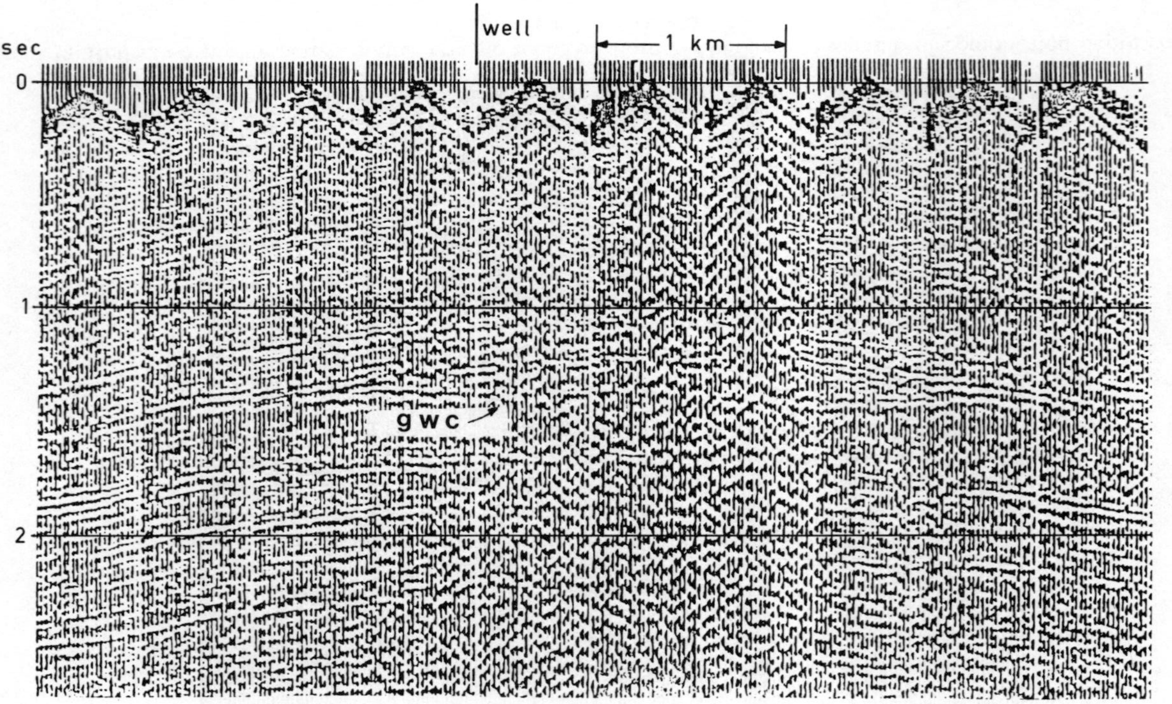

PROFILE Q. Early 1960s centre-spread profile across gas-bearing anticline showing reflection from gas–water contact, Pakistan.

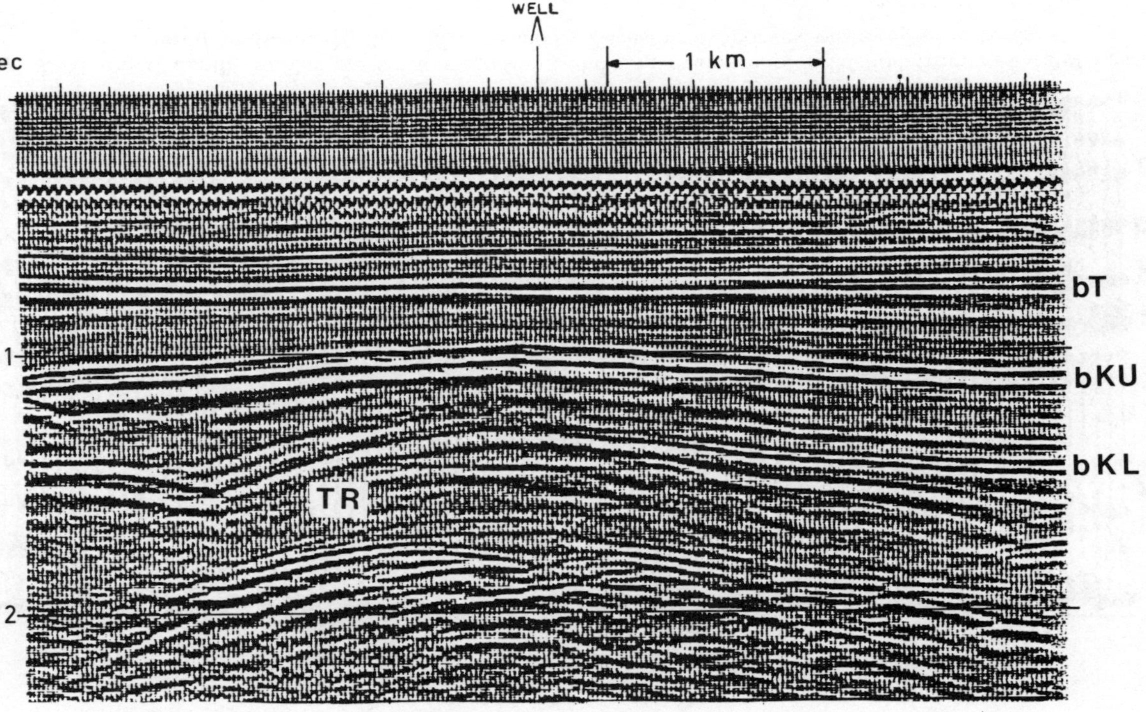

PROFILE R. Flat spot in Middle Bunter (Triassic) sandstone, offshore Netherlands.

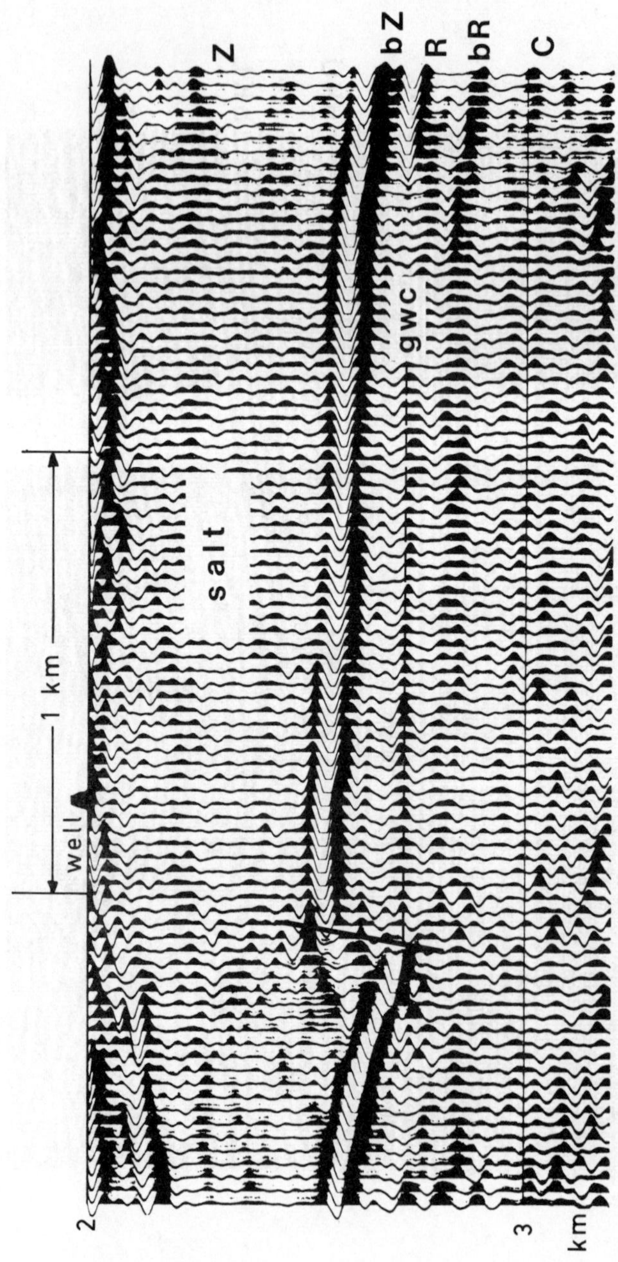

PROFILE S. Migrated depth profile (pre-stack migration), showing evidence of reflection from gas–water contact in Slochteren sandstone (Rotliegendes) sealed by Zechstein evaporites, onshore Netherlands.

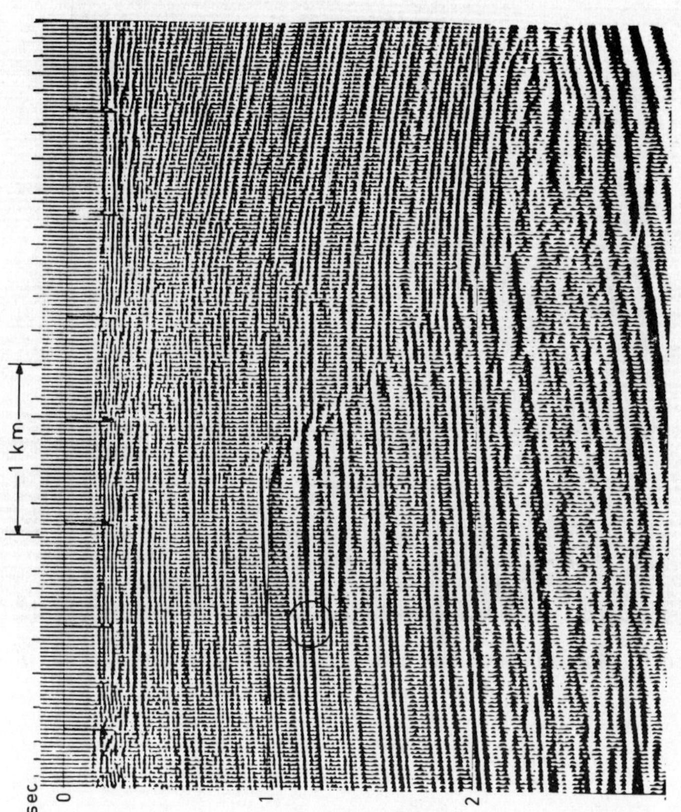

PROFILE T. Migrated profile showing gas accumulations in Tertiary sand section trapped by fault; example of bright spots and polarity inversion, offshore Brunei.

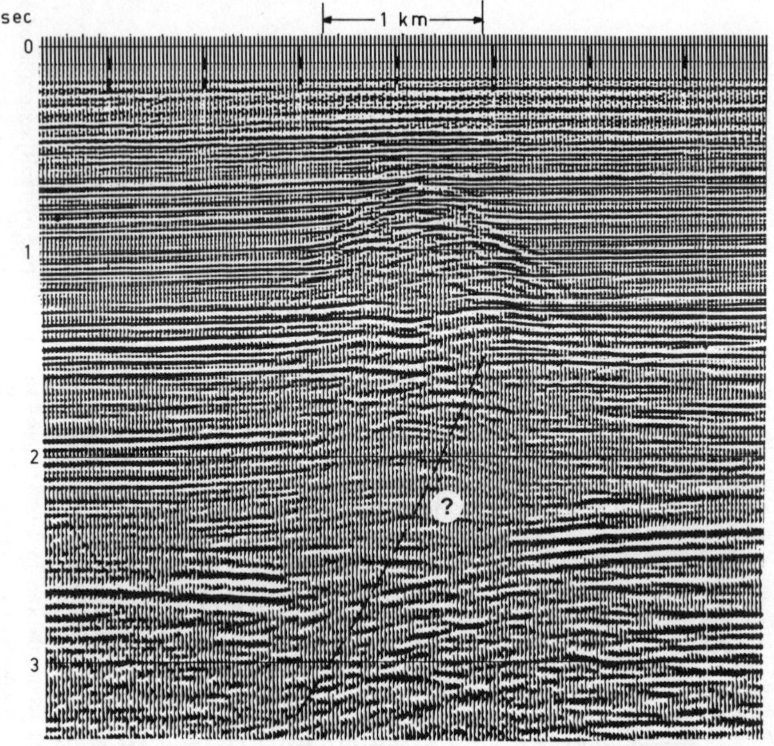

PROFILE U. Gas chimney and velocity anomaly in Tertiary section, offshore Brunei.

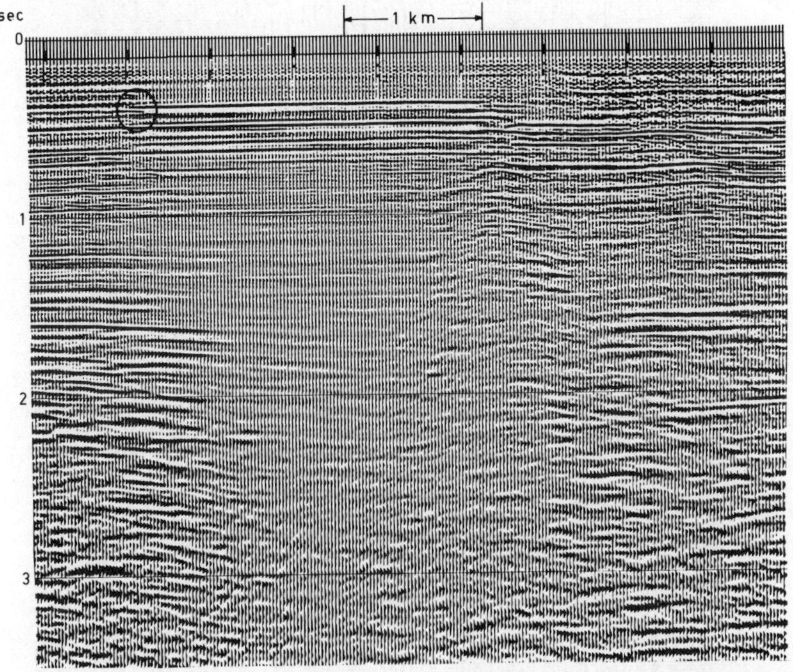

PROFILE V. Gas chimney, bright spots and polarity inversion in Tertiary section, offshore Brunei.

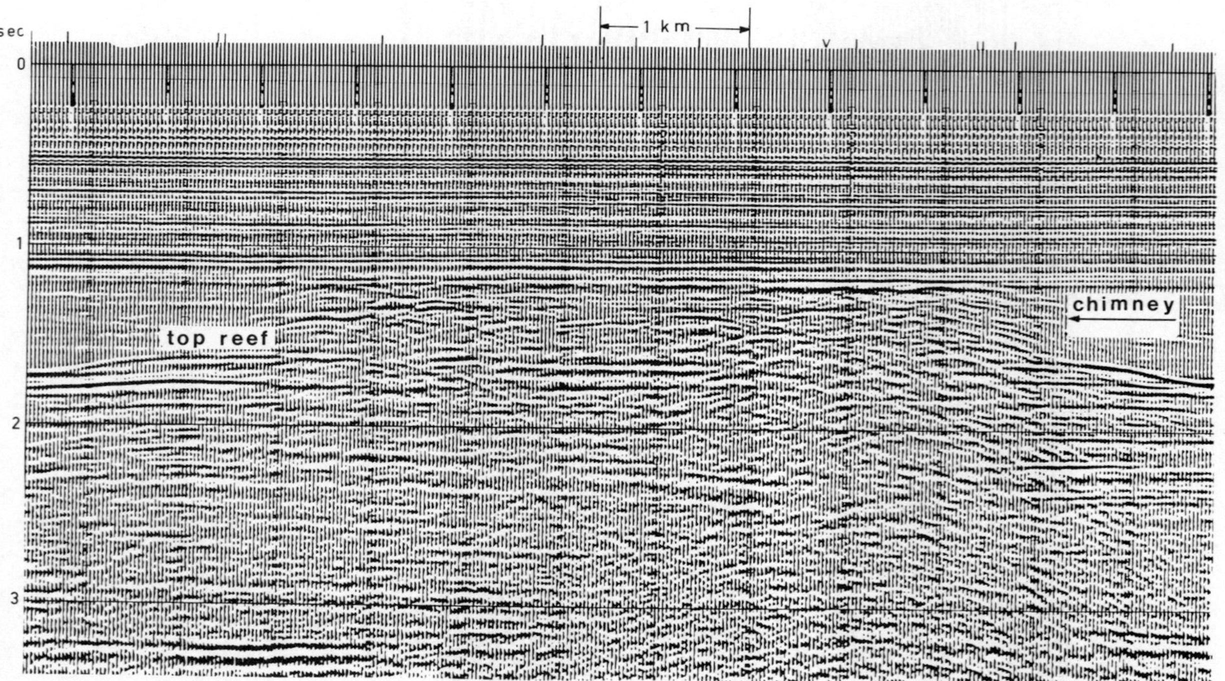

PROFILE W. Gas chimney in Tertiary section above gas-bearing limestone reef, offshore Sarawak.

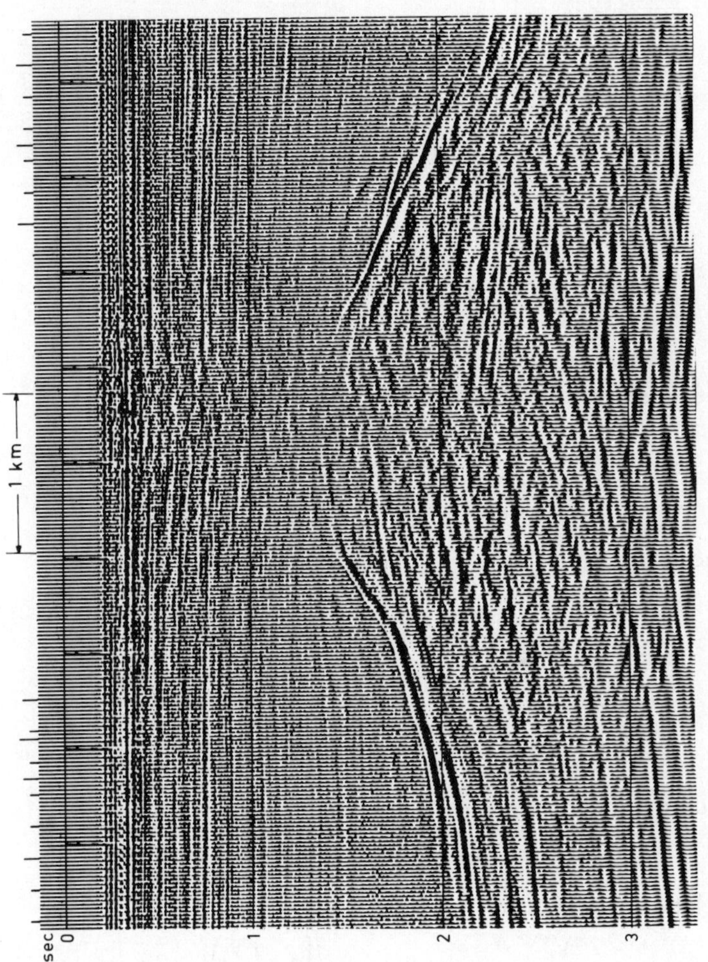

PROFILE X. Dim spot caused by gas accumulation in Tertiary limestone reef, offshore Sarawak, migrated profile.

Appendices

APPENDIX 1
DERIVATION OF RAY TRACING FORMULA OF SECTION 3.5

Solving the set of equations

$$\begin{aligned}1/V_i(n_3 i_2 - n_2 i_3) &= 1/V_t(n_3 t_2 - n_2 t_3)\\ 1/V_i(n_1 i_3 - n_3 i_1) &= 1/V_t(n_1 t_3 - n_3 t_1)\\ 1/V_i(n_2 i_1 - n_1 i_2) &= 1/V_t(n_2 t_1 - n_1 t_2)\end{aligned}$$

we find for t_1

$$\begin{aligned}t_1 = \frac{V_t}{V_i}&[i_1 - n_1(i_1 n_1 + i_2 n_2 + i_3 n_3)]\\ &\pm n_1 \sqrt{\left\{\left(\frac{V_t}{V_i}\right)^2 [(i_1 n_1 + i_2 n_2 + i_3 n_3)^2 - 1] + 1\right\}}\end{aligned}$$

Considering that

$$\cos\theta_i = -(i_1 n_1 + i_2 n_2 + i_3 n_3)$$

this relation simplifies to

$$t_1 = \frac{V_t}{V_i}(i_1 + n_1 \cos\theta_i) \pm n_1 \sqrt{\left[1 - \left(\frac{V_t}{V_i}\right)^2 \sin^2\theta_i\right]}$$

or

$$t_1 = \frac{V_t}{V_i} i_1 + \left(\frac{V_t}{V_i}\cos\theta_i \pm \cos\theta_t\right) n_1$$

since

$$\frac{V_t}{V_i} \sin \theta_i = \sin \theta_t$$

Similar expressions hold for t_2 and t_3; hereby the minus sign satisfies the condition that the dot product $\bar{\mathbf{I}} . \bar{\mathbf{T}}$ is negative, so that we have for the transmitted ray

$$t_j = \frac{V_t}{V_i} i_j + \left(\frac{V_t}{V_i} \cos \theta_i - \cos \theta_t \right) n_j \qquad \text{for } j = 1, 2, 3$$

or
$$\mathbf{T} = \frac{V_t}{V_i} \mathbf{I} + \left(\frac{V_t}{V_i} \cos \theta_i - \cos \theta_t \right) \mathbf{N}$$

APPENDIX 2
EVALUATION OF V_{nmo} IN HORIZONTAL VELOCITY DISTRIBUTION

From Section 3.8,

$$1/V_{\text{nmo}}^2 = f^{(1)}(x^2) = \frac{\mathrm{d}t_x}{\mathrm{d}x} \frac{t_x}{x} \qquad \text{for } x = 0$$

With reference to Fig. 3.30 and Section 3.6, it obtains that

$$\frac{\mathrm{d}t_x}{\mathrm{d}x} = \frac{\sin \theta_{\mathrm{r}}}{2V_0} - \frac{\sin \theta_{\mathrm{s}}}{2V_0}$$

whereby, from Snell's law,

$$\sin \theta_{\mathrm{r}} = (V_0/V_{\mathrm{R}}) \sin (i_{\mathrm{R}} + 2\delta)$$

$$\sin \theta_{\mathrm{s}} = -(V_0/V_{\mathrm{R}}) \sin i_{\mathrm{R}}$$

V_{R} is the velocity at the level of the reflection point.

Ray parameters for the downgoing and reflected rays are $p_{\mathrm{s}} = \sin \theta_{\mathrm{s}}/V_0$ and $p_{\mathrm{r}} = \sin \theta_{\mathrm{r}}/V_0$ respectively. Hence, from Section 3.4,

$$t_x = \int_0^z \frac{\mathrm{d}z}{V_z(1 - p_{\mathrm{s}}^2 V_z^2)^{\frac{1}{2}}} + \int_0^z \frac{\mathrm{d}z}{V_z(1 - p_{\mathrm{r}}^2 V_z^2)^{\frac{1}{2}}}$$

and
$$x = \int_0^z \frac{p_{\mathrm{s}} V_z}{(1 - p_{\mathrm{s}}^2 V_z^2)^{\frac{1}{2}}} \mathrm{d}z + \int_0^z \frac{p_{\mathrm{r}} V_z}{(1 - p_{\mathrm{r}}^2 V_z^2)^{\frac{1}{2}}} \mathrm{d}z$$

so that for $(\mathrm{d}t_x/\mathrm{d}x)(t_x/x)$ we can write

$$\frac{\mathrm{d}t_x}{\mathrm{d}x}\frac{t_x}{x}=\frac{p_r-p_s}{2}\,\frac{\displaystyle\int_0^z \frac{\mathrm{d}z}{V_z(1-p_s^2V_z^2)^{\frac{1}{2}}}+\int_0^z \frac{\mathrm{d}z}{V_z(1-p_r^2V_z^2)^{\frac{1}{2}}}}{\displaystyle\int_0^z \frac{p_sV_z}{(1-p_s^2V_z^2)^{\frac{1}{2}}}\mathrm{d}z+\int_0^z \frac{p_rV_z}{(1-p_r^2V_z^2)^{\frac{1}{2}}}\mathrm{d}z}$$

At the c.m.p. this expression is indeterminate, since both $\mathrm{d}t_x/\mathrm{d}x$ and x are then equal to zero. From L'Hôpital's rule its limiting value can be determined upon substitution of $i_R=-\delta$ after differentiating $\mathrm{d}t_x/\mathrm{d}x$ and x with respect to i_R. This will yield the V_{nmo} relation given in Section 3.8.

APPENDIX 3
INCLINED DIFFRACTION LINE SOURCE IN LINEAR VELOCITY DISTRIBUTION

With reference to Fig. A.1, let point $P(x, y)$ be a coincident source–detector location. Consider a wavefront sphere from P which is tangent to the line source in the X–Z plane, the point of tangency representing the diffraction point D. We shall denote the radius of this wavefront by R, and the depth of its centre by z_c.

From equation 3.9 we have for the two-way diffraction time at P

$$T(x, y)=\frac{2L}{V_0}\cosh^{-1}\left(\frac{z_c}{L}+1\right)\qquad\text{with } L=V_0/k$$

The value for z_c in the above expression is derived as follows. The intersection of the wavefront and the X–Z plane is a circle with radius $(R^2-y^2)^{\frac{1}{2}}$, from which

$$\cos\delta=\frac{(R^2-y^2)^{\frac{1}{2}}}{x\tan\delta-z_c}$$

so that

$$z_c=x\tan\delta-\frac{(R^2-y^2)^{\frac{1}{2}}}{\cos\delta}$$

From equations 3.8 and 3.9,

$$R^2=z_c^2+2Lz_c$$

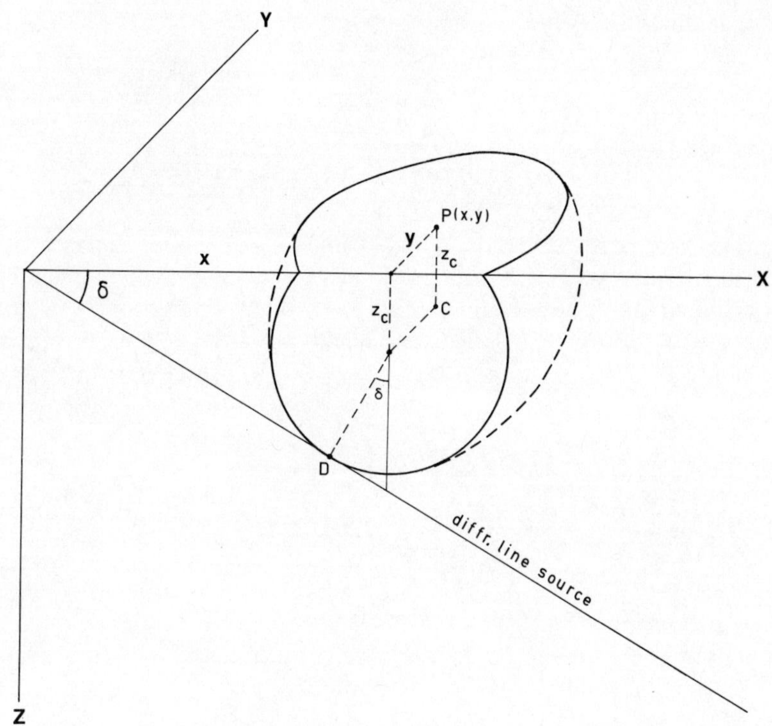

FIG. A.1. Three-dimensional diffraction geometry.

From these last two relations,

$$z_c = \frac{-(L+\frac{1}{2}x\sin 2\delta)+[(\frac{1}{2}x\sin 2\delta+L)^2+\sin^2\delta(x^2\sin^2\delta+y^2)]^{\frac{1}{2}}}{\sin^2\delta}$$

For the special case that $\delta = 90°$, we have

$$z_c = -L+(L^2+x^2+y^2)^{\frac{1}{2}}$$

Then, correspondingly,

$$T(x, y) = \frac{2L}{V_0}\cosh^{-1}\left(1+\frac{x^2+y^2}{L^2}\right)^{\frac{1}{2}}$$

or $$T(x, y) = \frac{2L}{V_0}\sinh^{-1}\frac{(x^2+y^2)^{\frac{1}{2}}}{L}$$

which represents the envelope of a set of maximum convexity surfaces.

Index